A History of Spaces

T0179151

A History of Spaces provides an essential insight into the practices and ideas of maps and map-making. It draws on a wide range of social theorists and theorists of maps and cartography to show how maps and map-making have shaped the spaces in which we live.

The book begins by asking a seemingly simple question: what does it mean to draw a line? It then gives a seemingly simple answer: to create a boundary, to define a space, and to shape an identity. The book builds on this foundation by exploring how, historically, maps have reached deep into social imaginaries to code the modern world. Going beyond the focus of traditional cartography, the book draws on examples of the use of maps from the sixteenth century to the present, including their role in projects of the national and colonial state, emergent capitalism and the planetary consciousness of the natural sciences. It also considers the use of maps for military purposes, maps that have coded modern conceptions of health, disease and social character, and maps of the transparent human body and the transparent earth.

The final chapters of the book turn to the rapid pace of change in mapping technologies, the forms of visualization and representation that are now possible, and what the author refers to as 'the possibilities for post-representational cartographies'.

John Pickles is Earl N. Phillips Distinguished Professor of International Studies and Professor of Geography at the University of North Carolina at Chapel Hill.

A History of Spaces

Cartographic reason, mapping and the geo-coded world

John Pickles

Routledge
Taylor & Francis Group

LONDON AND NEW YORK

First published 2004
by Routledge
2 Park Square, Milton Park, Abingdon, Oxon, OX14 4RN

Simultaneously published in the USA and Canada
by Routledge
711 Third Ave, New York NY 10017

Routledge is an imprint of the Taylor & Francis Group

Transferred to Digital Printing 2006

Typeset in Times by Wearset Ltd, Boldon, Tyne and Wear

British Library Cataloguing in Publication Data
A catalogue record for this book is available from the British Library

Library of Congress Cataloging in Publication Data
Pickles, John, 1960–
    A history of spaces : cartographic reason, mapping, and the geo-
coded world / John Pickles.
        p. cm.
    Includes bibliographical references and index.
    1. Cartography. I. Title.
    GA105.3.P52 2004
    526–dc21   2003008283

ISBN - 978 0 4151 4498 8

For Lynn, Leon and my parents, and for three teachers:
Roger Downs, Peter Gould and Joseph Kockelmans

One pylon marks the spot
BBC News Monday, 15 October 2001 11:55GMT,
http://news.bbc.co.uk/hi/english/uk/england/newsid_1600000/1600225.stm

A field in North Lincolnshire is the most featureless part of the UK, according to a new Ordnance Survey (OS) map.

The square kilometre on the outskirts of the village of Ousefleet, near Scunthorpe, has nothing in it except a single electricity pylon and some overhanging cable. Grid reference SE830220 on map 112 is as near as cartographers can find to a completely blank square among the 320,000 in the widely-used Landranger map series...

The quest to find Britain's most boring place was set by a listener to John Peel's *Home Truths* show on BBC Radio 4.

Philip Round from the OS said: 'We're not saying it's the dullest place in Britain. It might be the most fascinating place on earth but on our Landranger maps it has the least amount of information. No ditches, streams or buildings in it are shown on this particular scale of map. That's quite some going, considering the low-lying areas of East Anglia and remote parts of Scotland.'

'Friendly community'
The land has been farmed by the Ella family for over 100 years. Tom and Avril, both in their 50s, grow wheat, barley and sugar beet on it.

'It's a lovely place to live,' Mrs Ella said. 'It's a small, friendly community with a lovely church nearby.'

But the family is not impressed by the thought that map enthusiasts may soon be flocking to the location.

'If people want to come and look at a field, I don't mind, but they're wasting their petrol.'

The OS has warned that the field's claim to fame could be threatened by more detailed maps of the same area. 'Drainage ditches or dykes might be shown up,' said Mr Round.

Contents

Illustrations

Figures

Table

Preface and acknowledgements

PREFACE

This book began life with the working title of *Mapping and Social Theory* as one in a Frontiers in Human Geography series with the express intent of synthesizing for students and general readers the rapidly emerging role being played by social theories of one kind or another in reshaping and revisioning thematic areas of geographical thought and practice (see Agnew 1998). The book still does this, but it has also mutated, as social theory itself has, into a much more immanent reading of the practices and ideas of spatial thought, mapping and map-making. Throughout, I have drawn on a wide range of social theorists and theorists of maps and cartography in an effort to flesh out some of the many ways in which we can think of the ways in which cartographic reason has coded our world.

The title of the book captures precisely its content. 'A History of Spaces' refers to Michel Foucault's (1986) suggestion that 'a whole history of spaces has still to be written'. I have not attempted to write a whole history of spaces or a comprehensive genealogy of maps and mapping. Such a task is beyond the reach of a single book or author. The history of maps and mapping itself is a massive topic, as the ambitious History of Cartography Project – now in six volumes in twelve books with about 7,000,000 words – amply demonstrates. Instead I have tried to write *a* history of spaces. I take my starting point with social theories from Heidegger to Adorno and beyond by putting in question all representational epistemologies and logics. In this task, 'mapping' is my central concern. I draw on maps and mapping as my point of entry into a consideration of the ways in which 'Cartographic Reason' can – as Gunnar Olsson, Franco Farinelli and Tom Conley have variously suggested – be seen as the missing element in social theories of modernity. In using 'the Over-Coded World' I explicitly associate my reading of the role of mapping in shaping social, spatial and natural identities with Althusserian notions of over-determination, Gibson-Graham's arguments against essentialism, and a Deleuzian and Guattarian project of immanent materialism. The book points, above all, to the ways in which our lives have been and are being shaped and constituted through myriads of intersecting and overlapping mappings in use every day. Over-coding thus points to the ways in which the formation of identities can be seen, in part, as a kind of spatialized historical process of mapping occurring at many scales simultaneously. In this sense the book takes up Henri Lefebvre's (1991: 85) question: 'How many maps, in the descriptive or geographical sense, might be needed to deal exhaustively with a given space, to code and decode all its meanings and contents?'. He answers: 'It is doubtful whether a finite number can ever be given to this sort of question.' These multiple and overlapping inscriptions are also spaces of slippage in the process of identity creation, signalling possibilities for other readings and practices of mapping. The book concludes with these slippages and the opportunities for rereading maps and mapping they seem to provide.

A *History of Spaces* owes much to my varied scholarly collaborations over the past thirty years. I went to Penn State for graduate study in 1978 to work on what I perceived to be fundamental problems in behavioural geography and approaches to cognitive mapping. The theories of mind and subject that underpinned these approaches to mapping were deeply problematic to me at the time and have remained so ever since. Roger Downs and Peter Gould each encouraged this thinking, even if not always agreeing with it. Their probing interrogations helped enormously in my efforts to tease out a critique of the work to which, in many ways, they had devoted a large part of their professional lives. Peter Gould passed away in 1999, but he already knew where I was going. Indeed, he had travelled most of the paths before me. I am proud to have travelled with him a little along the way. In preparing his *The Geographer at Work* (1985) Peter often said that he wanted to write a book that he could give to non-geographers to answer the question often asked of geographers, 'So what do geographers do?'. In part, I have been motivated by a similar question. With the spatial turn in social theory, the social sciences and the humanities, cartographic and mapping metaphors have proliferated. But even such wonderful works as Geoff King's *Mapping Reality* (1996) have given only limited attention to the heritage of cartographic and geographic thought. In response, I have sought to write a book that unpacks that long heritage of engagement with cartographic reason in ways that opens a two-way conversation between geographers and non-geographers about the post-disciplinary sensibility wrought by these most recent spatial and cultural turns. More succinctly, I have asked, 'what do maps do?' and I have tried to show how 'maps matter!'.

The more I have worked on writing this book, the more I have encountered Roger Downs there before me. In exploring mapping and social theory I have come to realize more than ever the fundamental debt my thinking owes to Roger's books *Image and Environment: Cognitive Mapping and Spatial Behaviour* and *Maps in Mind: Reflections on Cognitive Mapping*, and to the hours of conversation in our advising meetings over those years. Roger's fascination with the emergence and development of spatial imagination among children and professionals alike has contributed substantially to stretching what counts as cartography to include popular cartographies of one kind or another. The experience of reading *Maps in Minds* remains fresh today.

When I first travelled to the United States, I carried with me one book to read on the aeroplane. It was *Phenomenology* (1967) edited by Joseph Kockelmans. At the time, it was a curiosity encountered in a bookstore in Manchester and I had no idea that Kockelmans was a professor of philosophy at Penn State. The focus of the book and my subsequent encounter with Joe has been vital in everything I have attempted since. Much of my work in philosophy and the humanities was supported through Interdisciplinary Fellowships in the Humanities administered by Joe and through philosophy courses, seminars, working groups and personal meetings with him. It is to him that I owe any reading skills I have been able to acquire. It is also through him that I was introduced to hermeneutic thought and from him that I gained a lasting appreciation of and love for continental philosophy. He was crucially important in introducing me to and guiding me through phenomenological and hermeneutical critiques of both naturalism and psychologism, and in so doing set me on a Husserlian and Heideggerian path to a 'post-structural materialism' that informs this work. Students often ask me, 'what happened to your interest in phenomenology?'; I hope here I have made it clear that I remain, unabashedly and deeply, a phenomenologist albeit of a hermeneutic and post-structural and post-Marxist kind. I hope it is also clear that the hermeneutic ontology of spatiality hinted at in the final chapter of *Phenomenology, Science and Geography* (1985) is given detailed and reworked form in this book.

I have learned from the example of these three individuals what it means to take texts seriously and to allow (and perhaps expect and demand of) others to push hard against such serious ideas and works. I have spent many years wandering in what might have appeared to my teachers to be different areas of geography, but I hope they would see in this work a return to that which was always present. It is to these three teachers that I have dedicated the book in the belief that what I have written in it is little more than a triangulation of my conversations with each of them.

In what follows I have tried to read across a wide range of works on mapping, from professional to popular forms, and from more traditional to more avant-garde works. I have tried to read these texts 'openly' in ways that seek to breath new life into them. Of late, I have become increasingly weary (and wary) of forms of dismissive text-reading that discard older works simply because of some supposed or stated epistemological or political commitment, or fail to find in those works challenges of rereading and re-placement. I have long been equally weary and wary of charges that critical theoretical work is difficult, even that it uses jargon (a charge those who make would rarely apply to scholarship in their own technical journals and books). Frankly, I remain deeply optimistic about the challenges of intellectual work and the political value of disciplinary and intellectual histories. I am particularly convinced of the importance of an ethics of reading that struggles with the text itself – especially in those cases where the text appears difficult, unclear or just plain wrong. I learned this ethics of reading watching Joe Kockelmans unpack even the most obscure phrasing in Husserl, Heidegger and Merleau-Ponty. I have come to better understand its political importance in reading Derrida, particularly in consort with Wolfgang Natter and other colleagues and visitors to the Committee on Social Theory at the University of Kentucky. In this book I have tried to put into practice this ethics of reading. I have attempted to reflect the opportunities for thinking the cartographic imagination in and through a wide range of readings and forms.

   I am sure also that many readers may feel this or that work should also have been considered in the pages ahead. This book was initially conceived and written for a general and student audience, with a corresponding editorial charge to minimize citations. The book has morphed beyond that original goal, but for those readers who do feel that more should have been included I can only respond with the hope that the selective analyses presented here of the social lives of maps contribute to the production of other histories of spaces and maps, to the recentring of attention on maps and mapping within disciplinary and interdisciplinary debates about spaces, and perhaps to a richer understanding of the social lives of maps in cartography itself.

   During the time I studied at Penn State, the Department of Geography was a hive of 'mapping' activity; graduate seminars entitled 'Maps and mapping', graduate-student working groups on mapping and spatial thought, and what seemed at the time like many different lines of flight through (and sometimes from) spatial analysis. During those years (1978–83) Roger Downs and David Stea had recently published *Maps in Minds,* Ron Abler was working hard to establish the Deasy Cartographic Laboratory, Peter Gould was opening up mathematical notions of relation, mapping and function as ways to explore new cartographies of multidimensional space, Peirce Lewis ran the departmental topographic map library for undergraduate and graduate students alike at every turn opening up the magical worlds of what he would later characterize as 'cartophilia' (Lewis 1985: 465), Wilbur Zelinsky continued to push the profession and students to think creatively about maps and mapping, Paul Simkins delighted everyone with new weekly productions of 'stained glass' choropleth maps, and Greg Knight's system-mappings brought the abstract possibilities of environmental analysis to life for many of us for the first time. Alan Rodgers, Will and Ruby Miller, Lucky Yapa, Rod Erickson and Fred Wernstedt rounded out the department and provided a model environment within which ideas could be engaged. In that environment Noriyuki Sugiura, Don Kunze, Kathy Christiansen, Jim Meyer, Mark Morey, Rich Schein, Jim Ackerman, Donna Shimamura, Doug Markey, David Black, Karen Schmelzkopf, Dan Baker, and many more were part of the broader conversation on mapping and the epistemologies, aesthetics, rhetoric and politics of representation. I have heard the echoes of their arguments at every turn in writing this book. I have continued to benefit from the ongoing fascination at Penn State with cartographic theory through the works and assistance of Alan MacEachren, Jeremy Crampton, John Krygier, Matt Hannah and Ulf Strohmeyer.

It was also while a graduate student at Penn State that Ron Abler, Roger Downs, Peter Gould and Peirce Lewis independently, introduced me to the 'underground' works of Bill Bunge. Particularly in his work on children's maps and the community of Fitzgerald, Bunge asked the intriguingly simple question: what geographies would we write and what worlds would we build if we mapped the experience of children and African-Americans in this world? Mapping the spaces of broken glass, torn fences and child injuries, Bunge showed us not only the real material spaces produced by our 'adult' worlds and the ways in which these spaces, cities and maps marginalized our own children, exposing them to danger at every turn, but he exposed in one stroke what had become clear to me through three years of work and engagement in South Africa – the truth claims of science must always be interrogated from the position from which you look, the shoes in which you walk and the maps you construct to guide your journeys. As Bunge and his expeditionary forces at the time were demonstrating, the techniques of science, analysis and representation are open and thoroughly contestable, available to us to use for various and different purposes. I refer to these as 'underground' works because that is exactly what they seemed to be in a discipline that was at that time so thoroughly captivated by various streams of structuralist, functionalist and generally reductionist logic. Bunge's work – apparently like the man himself – seemed to barge into the spaces of scholarly calm and dullness and shout loudly about the need for new thinking, a new positionality and new sensitivity in regard to the tools with which we map social and natural worlds so thoroughly inscribed by patriarchy, capitalism, nationalism and militarism. And he did it, as Paolo Freire had taught us, through a process of *conscientization* – a careful mediation of technical skills, people's education and cultural politics in which the 'giving of language and the ability to speak' is an important part of the struggle for social justice for the marginalized.

In 1983, I was able to teach for a semester at the University of Minnesota and was asked to 'fill in' for Fred Luckerman's course 'History of geographic thought' and Yi-Fu Tuan's 'Space and place'. I realized only recently how fundamental this experience has been to writing this book when rereading Samuel Edgerton's *The Renaissance Rediscovery of Linear Perspective* (Basic Books 1975) and John White's *The Birth and Rebirth of*

Pictorial Space (Boston Book and Art Shop 1967). I had used these texts for those courses in trying to come to grips with the centrality of perspective in spatial theory. In rereading these works, I have realized how much of this material I had already worked through and how influential it had become in my readings of spatiality and visuality in the discipline. I am grateful to Fred and Yi-Fu for being on assignment that semester and for graciously allowing a new Ph.D. to take over their well-established courses. The chance to teach those two courses forced me out of the critique of metaphysics, liberal humanism and spatial science that grounded my work at Penn State into an analysis of the cultural practices of representation and vision that seemed to me to underpin the logics and metaphors of space in the discipline of geography. The adventure of reading and teaching about visual space, pictorial space and the spaces of experience has, however, only now found any concrete outlet. During that time I had the privilege to participate in many long conversations variously with John Adams, Roger Miller, Joe Schwartzberg, Eric Sheppard and other faculty members and graduate students such as Trevor Barnes, Michael Curry, Patrick McGreevy and April Veness about issues of intellectual history, spatial thinking and the then newly emerging critique of cartographic reason.

Over the years, I have been privileged to know quite a few 'craft' cartographers schooled in Central European cartography, in particular Bruno Martin, Hubertus Bloemer and Gyula Pauer. I remember watching with fascination as Bruno Martin and Raymond Poonsamy created maps in the office next to mine in the Department of Geography at the University of Natal in Pietermaritzburg. Pouring over the mapping tables and working in the darkroom for hours each day (a concentration broken only by cigarette and tea breaks), Bruno and Raymond produced exquisitely detailed and subtle maps of immense rhetorical power in a society whose own outlines were drawn in black and white, whose borders were being 'cleaned up' (through gerrymandering, forced removals and assassinations), and whose populations knew only too well how lines on maps shape daily life. What puzzled me then, and still does, was how maps of such clarity and power are constructed out of such simple earthen materials and yet function in a society to such devastating and powerful effect. How did the tools of the desk and darkroom meld in the hands of the skilled cartographer into graven images of such power? How did such simple lines on the map signify – in the hands of the apartheid state – such terrifying projects of social engineering? And how could we use these maps to disrupt the terrible inscription of boundaries between people and places that so typified apartheid South Africa (a topic that exercised us all at the time)?

Later, during visits to Hungary and Bulgaria, I encountered the Central and East European cartographic traditions of spectacular national atlases. Their crafty magic still fills me with awe for its precision and detail, even as it reminds me of the universalizing goals of state cartographies. At West

Virginia University I encountered another form of state cartography through the painstaking archival reconstruction of electoral boundaries in the historical cartography of Ken Martis. This task of spatial exegesis and reconstruction seemed at the time, and still seems, to be a different kind of mystical world-making, an act I encountered again recently in the historical cartographic exegesis of the research institute of the Dipartimento di Scienze Geografiche e Storiche at the University of Trieste.

In the US, I have been privileged to see the emergence of a new generation of 'craft' cartography, and to see how the magic of map-making works at close hand in the mappings of Hubertus Bloemer, Gyula Pauer and Dick Gilbreath. Each of these has taught me more than they realize about cartography and a good deal more about the ways in which maps are complexly crafted texts, and each has inducted me in various ways into the joyful mysteries and magical practices of map-making.

As a result of these encounters and fascinations, the post-empiricist discussion of the crisis of representation has always meant for me something that is both abstract and practical. For the past ten years I have been carrying out research in Bulgaria, and before that in South Africa. Each setting has instilled in me a fascination for the political functioning of maps. In South Africa, territorial identities and the social lives of millions were imprinted by fiat by the constant tinkering of apartheid technocrats to satisfy this or that interest and this or that racist aesthetic. For the people with whom I worked this crisis of representation was very much a crisis of having their lands and lives mapped, rationalized and reordered by the forces of racial nationalism and racial capitalism. In Bulgaria it has not been the sheer force of the mapping project and the inscription spatialized identities that has impressed me, but the absence of maps – the sheer inability to obtain maps under a regime of intense secrecy about the mapped image and mapped landscape. The corollary to this absence has been the stunning revelation that my own colleagues, whose professional lives have been built on the collection and representation of spatial data of one sort or another, were themselves utterly perplexed about how to construct maps in the face of this basic absence of state-supplied geographical information. Apparently, even for professional geographers the absence of the topographic map can be a fundamental barrier to any mapping. In negotiating this particular crisis of representation – the secret lives of maps for my Bulgarian colleagues and research lives without maps for my American colleagues – I am indebted to Bob Begg, Rumiana Dobrina, Jim Friedberg, Kristo Ganev, Boian Koulov, Didi Mikhova, Zoya Mateeva, Mariana Nikolova, Krassimira Paskaleva, Phil Shapiro, Angel Sharenkov, Stefan Velev, Brent Yarnal, the John D. and Catherine T. MacArthur Foundation and the National Science Foundation.

In 1990, I met Brian Harley at the Annual Conference of the Association of American Geographers. After our session together, we went to drink beer and talk about what we took to be an emerging uncritical valorization of geographical information systems (GIS). The outgrowth of that 'chat' was *Ground Truth: The Social Implications of Geographic Information Systems* (1995), completed after Brian's death on 20 December 1991. Subsequent to its publication, I have been involved in a series of fascinating and rewarding engagements with practitioners and theorists of GIS in various settings. Among them are Nick Chrisman, Helen Couclelis, Jeremy Crampton, Michael Curry, Matthew Edney, Greg Elmes, Michael Goodchild, Jon Goss, Trevor Harris, Francis Harvey, Ken Hillis, John Krygier, Helga Leitner, David Mark, Patrick McHaffie, Bob McMaster, Tim Nyerges, Eric Sheppard, Dalia Varenka, Howard Veregin and Daniel Weiner. These individuals have greatly expanded my own thinking about the mapping process and, as well, have forced me to think more seriously about the materiality of mapping technologies and practices. I was introduced to the institutional world of maps and map libraries, particularly at the Map Division of the Library of Congress and the British Map Library at the planning meeting for Volume 6 of the History of Cartography Project: *Cartography in the Twentieth Century* (courtesy of David Woodward and Mark Monmonier). It was here, and through subsequent meetings at the Library of Congress and with Jim Ackerman at the Newberry Library in Chicago, that my initial fascination with maps and mapping practices was extended to the technics and pleasures – as well as the questions – of map archives.

For the decade of the 1990s, the Department of Geography and the Committee on Social Theory at the University of Kentucky were my home for collective engagement with the spatial turn in social thought in all manner of forms. Paola Bacchetta, Dwight Billings, Stan Brunn, Andy Grimes, John Paul Jones, Michael Kennedy, Wolfgang Natter, Karl Raitz, Herb Reid, Sue Roberts, Ted Schatzki, Rich Schein, Karen Tice and Dick Ulack in particular have each contributed variously to the ideas in this book. The book has been influenced by the many visitors to the Committee on Social Theory seminars and colloquia. Ben Agger, Russell Berman, James Boon, Sam Bowles, Susan Buck-Morss, Judith Butler, Stewart Clegg, Arturo Escobar, Gustavo Esteva, Herb Gintis, Peter Jackson, Martin Jay, Doug Kellner, David Harvey, David Hoy, Thomas Laqueur, Charles Lemert, David Lloyd, Emily Martin, Doreen Massey, Timothy Mitchell, Gunnar Olsson, Steve Pile, Michael Roth, Bonnie Smith, Charles Tilly, Michele Wallace, Sam Weber, Iris Young, and the many other guests and family of social theory at UK have each, unwittingly, left their mark on this manuscript. I am particularly indebted to Wolfgang Natter, John Paul Jones and Ted Schatzki for long hours and many years of comradely conversations about critical social theory. Many graduate students have also influenced the shape and arguments of this book as we worked together through text after text in my seminars on disciplinary history and research design. To them all I am deeply grateful for their openness and excitement in addressing difficult texts and complex issues. Several graduate students, in particular, assisted with specific parts of the book. In particular, I would like to thank Keiron Bailey, Carl Dahlman, Michael Dorn, Owen Dwyer, Eugene McCann, Matt McCourt and Josh Lepawsky. Along the way, other colleagues have stimulated ideas represented here or kindly provided me with information and feedback about parts of the work. In particular, I would like to thank Trevor Barnes, Gianfranco Battisti, Bob Begg, Michael Curry, Matthew Edney, Kathy Gibson, Julie Graham, Alan Pred, Mary-Beth Pudup, Eric Sheppard, Adrian Smith, Jenny Robinson, Michael Watts, Dan Weiner and David Woodward.

Through the Tours conference on space and mapping, I came to understand more of the insightful work of Denis Cosgrove and Ola Soderstrom. In particular, their commitments to understanding the historical formations of cartographic and mapping practice have left their mark on this work. Denis has, in many ways and in different forms, already written this book. Ola – more than anyone I know, and certainly more than I – should have done so. At various other times, Gunnar Olsson, Alan Pred, Dagmar Reichert, Nigel Thrift, Michael Watts and Benno Werlen have each created 'folds' in the fabric of my thought in ways that leave deep marks in this work.

The book would not have been written without the suggestion first coming from Derek Gregory and Linda McDowell. Derek has been an encouraging and supportive colleague since our first meeting in 1983 in his rooms in Cambridge. Several years ago, Trevor Barnes asked me to review Derek's *Geographical Imaginations* for *Environment and Planning A*. To my great embarrassment, I never did get the review written. But, in rereading the first three sections of *Geographical Imaginations* after completing this book, I realize how that review morphed and has emerged here; my debt in this book to Derek's *Geographical Imaginations* should be obvious to all.

Ann Michael of Routledge has been, again, a model of patience in putting up with missed deadlines. That patience allowed me to revisit in greater depth the writing of Gunnar Olsson and to work more closely with what has seemed like a flood of new texts on cartographic reason and practice from David Harvey, Denis Cosgrove, Matthew Edney and Tim Conley among others. I hope the time invested has allowed this book to move from more limited readings of cartographic practice to one more attentive to both the theoretical and practical currents of contemporary social theory. Melanie Attridge and Andrew Mould at Routledge assisted greatly in bringing the book to completion.

I have drawn on and reworked my previously presented and published work throughout. Earlier versions of my arguments were presented at conferences, particularly in 1998 at the North American Cartographic Information Society (NACIS) Conference in Lexington, KY, the 19th International Cartographic Conference in Ottawa, and the conference 'Speaking, writing, drawing space' at the Université François-Rabelais in

Tours, France, December 1998. Chapter 2 draws on 'Hermeneutics and Propaganda Maps'. I am grateful to Trevor Barnes, Hubertus Bloemer, Ruth Rowles, Henry Ruf and Michael Watts for their invaluable comments on earlier drafts of that paper. The manuscript has emerged out of, and has been informed by, parallel projects supported by the National Center for Geographical Information Analysis at Santa Barbara and Buffalo: Initiative 19: 'GIS and Society' and the Critical History of GIS Project. Chapters 8 and 9 rework and extend arguments I initially published in *Ground Truth*, and were presented to the Committee on Social Theory Work-in-Progress Series at the University of Kentucky.

I am deeply indebted to the following organizations for their financial support for research and writing related to this book. The University of Kentucky has supported the research and writing through travel and research awards. The National Center for Geographic Information Analysis at Santa Barbara and Buffalo supported research and writing on digital mapping and geographical information systems. John D. and Catherine T. MacArthur Foundation and the National Science Foundation (Geography and International Programs) awards supported research that led to writings on geographical information systems in Bulgaria. The National Endowment for the Humanities supported archival research in Munich and Leipzig. The Fulbright Commisssion for Italy and the Dipartimento di Scienze Geografiche e Storiche at the University of Trieste provided support for research and writing in Trieste.

I completed this work three times; first in winter 1997, again in winter 2001, then in winter 2002–3, each time during time granted to me to write and work every day by the kindness and understanding of my wife Lynn and the tolerance (at least most days) of my son Leon. That my parents have also ceded this time to me first for many years before, and in each of these years during their visits for the holidays I deeply appreciate. This final version is now being completed exactly four years later, a delay in part for which I have Gunnar Olsson to thank. The cog he threw in my wheel with his reflections on cartographic reason has been productive beyond words.

Perhaps because my writing on mapping has always been so difficult and drawn out, and because publication deadlines have been so delayed and extended, writing on mapping (be it on propaganda maps and hermeneutics, on geographical information systems, on state socialist control of spatial data or on mapping and social theory) has for me always been a journey accompanied by the loss of dear friends: Michael, Brian, Ken, Velma, Peter, Simphiwe, Ernest, among others. I hope my writing only adds to, and does not detract from, the world that they tried to build.

ACKNOWLEDGEMENTS

The author and publishers would like to thank the following for granting permission to reproduce material in this work.
Map and Geography Division, Library of Congress, Washington, DC: 1.1 Hispaniola. C. Columbus; 1.3 Henricus Hondius world map, 1633: 2.4 Roosevelt map of the US; 2.6 Albert Speier's map for the reconstruction of Berlin; 7.1 Bird's eye view of Phoenix; 7.3 Fire Insurance Map, Tombstone, Arizona (New York: Sanborn Map and Publishing Company, 1886).
Tate, London 2002: 1.2 *Newton* by William Blake.
Stadelsches Kunstinstitut und Galerie, Frankfurt am Main: 1.5 *The Geographer*, Vermeer (oil on canvas, 1668–9).
7-year-old Leon: 1.6 'Four square'.
The University of Chicago Press: 1.7 Marshal Islands stick chart (Source: Turnbull, 1993, *Maps are Territories*).
Jo Gould: 1.8 Relation ... Mapping ... Function. Peter Gould (*with permission Jo Gould*); 4.1 Peter Gould's surjective, bijective and injective mappings; 8.8 Peter Gould's mappings.
The Belgium American Educational Foundation: 2.2 Map-poster dropped by German aeroplanes to Allied troops in Belgium while they were fighting, about 25 May 1940. (Source: The Belgian Campaign and the Surrender of the Belgian Army. The Belgium American Educational Foundation, New York, 1940).
Museum of Modern Art, New York, ArtResource and Artists Rights Society: 2.7 *Dada Movement*, Francis Picabia (1919).
NASA: 4.2 Whole Earth, Apollo 17; 8.9 Vegetation Canopy Lidar Mission (http://essp.gsfc.nasa.gov/vcl/).
Nebraska State Historical Society Photograph Collections, Lincoln: 4.3 The Surveyor: Robert Harvey.
Musée des Beaux-Arts Lille: 5.3 Seventeenth-century perspectival view of Maastricht by Gravure Sollain. (Source: Plans en Relief: Villes Fortes des Anciens Pays-Bas Français au XVIIcS (1989)); 5.4 Platted spaces of the perfect state. Lille avant les traveaux de vauban. Gravure de Blaeu (1649) (Source: Plans en Relief: Villes Fortes des Anciens Pays-Bas Français au XVIIcS (1989)); 5.5 Citadelle de Tournay (top) and photograph of 'Le Plan en Relief de 1701, Tournay' (bottom). (Source: Plans en Relief: Villes Fortes des Anciens Pays-Bas Français au XVIIcS (1989)).
Oxford University Press: 6.3 Gridded lands: lines, landholdings, landscapes. From Hildegard Binder Johnson, *Order Upon the Land.*
American Society for Photogrammetry and Remote Sensing: The Imaging and Geospatial Information Society: 7.4 Award-winning map: Map of the village of Buc, Versailles, at the scale of 1:2,000 produced from photographs in 1861. The map won a gold medal in 1863 in Madrid. (Source: Blachut and Burkhard (1989)).

New Scientist, Harcourt, London: 8.3 'Desert Storm's Satellites of War'. Cover *New Scientist* 27 July 1991, No. 1179.
UbiSoft Entertainment/Mindscape: 8.3 'Cyber-Empires' ad by Strategic Simulations Inc. in *Computer Game Review* October 1992, Volume 2, Issue 3.
The National Library of Medicine: 8.6 The Visible Human Project. (http://www.nlm.nih.gov/research/visible/visible_human.html).
GeoTech Media (www.geoplace.com): 8.11 'The Art of Prospecting for Customers'

Part I

Introduction

My objective ... has been to create a history of the different modes by which, in our culture, human beings are made subjects.

(Foucault, *The Subject and Power*)

Part 1

Introduction

1   Maps and worlds

It has always been this way with the map-makers: from their first scratches
on the cave wall to show the migration patterns of the herds, they have
traced lines and lived inside them.

(Sonenberg, *Cartographies*)

In a presentation nearly a decade ago, the Swedish geographer Gunnar
Olsson began with the comments: 'I have come a long way to tell you a
story ... it is a story of a finger and an eye.'[1] He then asked: 'What is geography if it is not the drawing and interpreting of a line?'. And what is the
drawing of a line if it is not also the creation of new objects? Which lines
we draw, how we draw them, the effects they have, and how they change
are crucial questions (see also Olsson 1992a, 1992b, 1998).

In the years since Olsson's presentation, I have struggled with the
implications of his suggestion that the drawing of lines is a fundamentally
geographical and spatial act in which identities are 'inscribed' and the
logos of western thought is founded. Gunnar's questions, caught as
they are between a deep spatial analytic sensibility and a rethinking
of modernist epistemologies and practices, result in a rich geopolitics of
lines, boundaries and limits in which the geographical imagination
is pushed to what he called 'the dematerialized point of abstractness'.
Here I want to focus on the implications of his work for how we understand the cartographic project and the geographical imagination that
undergirds it, how – as the Sonenberg and Foucault quotations with which
I began suggest – we have lived within the lines we have traced and been
made the subjects we have become. Let me begin with Olsson's finger
and eye.

The finger is literally indexical – it points to something to draw our
attention to it. It literally fingers the flux of the world to identify something: it delimits from a field a point, a place, an object for our attention. It
stabilizes a particular meaning within a world of possible meanings. And in
the modern world it generally does this by asking us to look at this thing,
this object, this place. How, then, do we point to the world and which

'lines of power' are inscribed through our contemporary social and scientific practices? How do we map the world?

Olsson suggests a historical and analytical reading of three 'lines of power' at work in the drawing and reading of a line ('=', '/', '$\frac{S}{s}$'). Each is indicative of a particular epistemology and mapping. The first are the lines of equivalence ('='), an epistemology of realism in which the concept refers to the thing (what Richard Rorty called philosophy as the mirror of nature). The second is the slash, the line of relation, the dialectic ('/'), which – he asserts perhaps mischievously – we now know to be dead. The third line is the dash ('$\frac{S}{s}$') – the line that binds signifier and signified, and whose form is semiotics. Each represents for Olsson a fundamentally different epistemology, and the move from lines of equivalence to relation to signification also represents a transition in structure and practices of knowledge production: it reflects a fundamental reworking of the categories and institutions of disciplined practices.

In his 1987 introduction to *Institution and Interpretation* Sam Weber had similarly argued that an epistemic shift was underway. Drawing on the French philosopher Gaston Bachelard, Weber asked how traditional modes of thinking were being reconfigured within the sciences, what Bachelard had called 'The New Scientific Spirit'. At stake in these reconfigurations 'is nothing less than the *idea* or *ideal* of *knowledge* based on a notion of truth conceived in terms of the *adequatio intellectus et rei*' (the adequation of thought and thing), the sign of equivalence. As Weber goes on to explain:

> The effects of such problematization ... extend far beyond the domain of 'methodologies'. The widespread 'identity crisis' that is affecting a variety of different disciplines today is only the most obvious indication of a process of rethinking ... what has changed is the relation of identity to nonidentity, of inclusion to exclusion.... In short, the traditional conception that holds space and time to be measurable in terms of 'the point and the instant' is irrevocably shaken by contemporary science.
>
> (Weber 1987: x, xi)

This is precisely what Gunnar Olsson sought to capture in his 'lines of power'. But to the lines, Olsson added the 'eye'. The ways in which the world has been represented visually have, historically, been important elements of the ways in which we have come to understand and act upon the world. Crucial to this way of seeing the world has been the project of universal science and the emergence of what Marie Louise Pratt (1992) has called 'planetary consciousness'. Mapping and cartography – the drawing of lines and the bounding of objects – have been at the heart of this consciousness. Mapping technologies and practices have been crucial to the emergence of modern 'views of the world', Enlightenment sensibilities and contemporary modernities. The world has literally been made, domesticated and ordered by drawing lines, distinctions, taxonomies and hier-archies: Europe and its others, West and non-West, or people with history and people without history. Through their gaze, gridding, and architec-tures the sciences have spatialized and produced the world we inhabit. And, indeed, this is perhaps the crucial issue: maps provide the very con-ditions of possibility for the worlds we inhabit and the subjects we become.

Olsson's comments, therefore, point not only to an understanding of geography engaged with questions of identity and difference, with bound-aries and transformations, and with categories and their dissolution. But he points also to an understanding of a world that has, in large part, been made as a *geo-coded* world; a world where boundary objects have been inscribed, literally written on the surface of the earth and coded by layer upon layer of lines drawn on paper. Cartographic institutions and practices have coded, decoded and recoded planetary, national and social spaces. They have literally and figuratively over-coded and overdetermined the worlds in which we live. They have respaced the geo-body. Maps and map-pings precede the territory they 'represent'. Just as scientific facts are pro-duced through the overlay and repetition of circulating references (Latour 1999), so also 'a geographical discovery is not really made until it has been recorded with sufficient accuracy so that it can be visited again' (Thrower 1976: 659 quoting Skelton 1967: 4) (Figure 1.1): territories are produced by the overlaying of inscriptions we call mappings.

Each of Olsson's 'lines of power' can be read as descriptions of forms of lived experience. The drawing and reading of a line, the historical emer-gence of cartographic reason, the production and circulation of a map, and lived experience are so thoroughly and historically intertwined and over-determined. Take the example of my 7-year-old son poring over a new book on whales. On turning a page, he comes across a simple world map of the distribution of breeding and feeding grounds. In his first 'reading', his index finger slides across the page guiding his eye across the map. He locates feeding and breeding grounds and he understands somehow that he can read each relationally. Connected by red arrows indicating annual migration streams, the locations are now read dynamically as part of a

Figure 1.1 What is Geography if it is not the Drawing and Interpreting of a
Line? *Hispaniola*, C. Columbus (Library of Congress, Washington, DC)

seasonal flux in the lives of whales generally. It is clear to any observer, not just a father, that what is occurring in this discussion of fixed and relative location is a rich evocation of, and dwelling in, a world of symbolic and metaphorical forms. The map is literally 'swimming' with places, relations, fluxes, meanings and potentialities – with whales, waters and seasons – and perhaps particularly so for a 7-year-old. Already socialized into the world of abstract spaces (a surprise to his parents), he is able to 'read' the map with ease and joy; he dwells in the geographical worlds of whales for a few precious moments and through the map he structures his understanding of their worlds and he structures his own world a little more.

As Olsson indicated by adding 'the eye' to his story, modern science rests on what Derek Gregory (1994: 15) has called the *'problematic of visualization'*. From Descartes to Goethe, the experience of the healthy corporal eye was a direct and true reflection of reality (Crary 1995: 97–8). But, as Jonathon Crary (1995: 9) has suggested, such truth effects 'were, in fact, based on a radical abstraction', an epistemology of 'plain vision' (and the practices, instruments, and institutions that were associated with it) that naturalized sight as a source of clear unmediated knowledge (Krygier 1997: 30; see also Edney 1997). For geographers, the ways in which the map became 'a theory which geographers ... accepted' (Ullman 1953: 57) is the story of the radical abstraction of the practices of the finger and the eye, the history of the technologies and institutions of map-making and map use, and ways of seeing and thinking; a story we need to revisit (Figure 1.2).

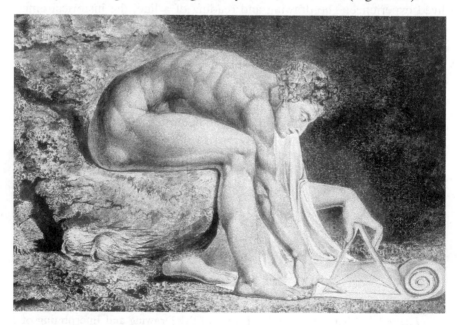

Figure 1.2 'The finger and the eye', *Newton* by William Blake (© Tate, London 2002, with permission)

In a broader sense, this is what the philosopher Martin Heidegger (1982) meant when he described the emergence of the 'age of the world-as-picture'. In using this phrase, Heidegger pointed to both the representational and objectifying nature of modern sciences *and* the global scope of the modern project; a rendering of all aspects of the world as picture, as 'standing reserve' or as resource for appropriation and use. Rendering the diversity of global alterities as objects, even commodities, for display and exhibition the global (European) project of modernity 'orders' and domesticates the unknown and the invisible, making them known and visible, making them available for use (Mitchell 1991). And in this sense Derek Gregory (1994) asks us to think about geographical practices such as mapping in terms of a broader epistemology and a politics that treat the 'world-as-exhibition'. The world-as-picture and as-exhibition was, in part, produced by technologies and practices of representation, including cycles of mapping, each of which left their residual impress on contemporary ways of seeing: the geometrical experiments of perspective; the exploratory portolan charts and the deep cultural fascination with boundaries (coastlines) that gave rise to them; the parcelling of land in the regional and national cadastres; the national topographic mapping programmes; the emergence of the globe as a cultural icon; and the more recent remote remapping of all aspects of social life (Figure 1.3).

Such a geopolitics of representation has very much been about property and the ownership and trading of commodities. As Walter Benjamin (1999) has shown in his writings on social and urban life in nineteenth-century

Figure 1.3 The World-as-Picture, Henricus Hondius world map, 1633 (Library of Congress, Washington, DC)

Paris, representation entered fully into the commodity relation by its production of an economy of display in which the spaces of the city were
restructured as spaces of visual display and mass consumption. The visual,
informational and the exotic were commodified for bourgeois consumption through the ur-forms of a new visual and global imaginary: the
national exhibition (Crystal Palace), the panorama, the plate-glass
window, and the shopping arcade in which the world of people, places and
goods were gathered for display and consumption (see Buck-Morss 1989).[2]

Maps and mapping have been at the heart of this economy of display
and demarcation (Figure 1.4). The globe has long served as an icon for
expansive capitalism and nationalism, and its iconic function continues to
inform representations of geographical reach, speed and power (see Cosgrove 1989, 1999, 2001). Moreover, geographers have long been fascinated
by the commodity form and have long pondered their unique and what Jan
Broek (1965: 64) referred to as their 'intimate' relationship to maps.[3] From
the descriptive mapping of commodity production and circulation in traditional regional and commercial geographies to the production of atlases,

Figure 1.4 The Plum Pudding in Danger; or, State Epicures Taking un Petit Souper,
by James Gillray, 26 February 1805. The cartoon depicts the rivalries
over territory between Napoleon and Pitt. Each is seated with the globe
served up on a platter like a plum pudding. Napoleon's sword has sliced
off France, Holland, Spain, Italy and Prussia, while his fork is dug spite-
fully into Hanover (then an appendage of the British crown). Pitt's fork
(trident) is stuck in the ocean as he carves the globe down the middle

globes and charts for display and sale, geography has been shaped by maps
and mapping. As Zelinsky (1973: 1) eulogized: 'surely we geographers and
map librarians, who are so helplessly smitten with map-love, know and
understand the objects we cherish, those beautiful, fascinating things that
are so much a part of our working lives and inmost thoughts and feelings'.

For both geographers and cartographers, the map has served in various
roles; as archive for geo-referenced data, as picture of the spatial order of
the world, as tool for investigating spatial relations, and as an object of
aesthetic and historical interest. The map has long been an object of fascination and value and is, for many geographers, the first picture they
choose to decorate their offices and their books. In a deep professional
and intellectual sense, the geographical imagination is one that is historically and personally identified with the cartographic image (Figure 1.5).

It is the drawing and interpreting of a line that marks the cartographic
impulse. What does it mean to draw and interpret a line, to make and use a
map, to dwell in the cartographic imagination? What are we to make of a
map such as that in Figure 1.6, drawn by my 7-year-old son to explain the
rules and strategies of the newly learned game of Four Square? In what ways
have mapping skills, abstract reasoning and spatial thinking so come to shape
the world of a 7-year-old, that the creation and use of such a map as a device
of complex and detailed explanation can already be such 'second nature'?

Figure 1.5 The Geographer, Vermeer (oil on canvas, 1668–9) (Stadelsches
Kunstinstitut und Galerie, Frankfurt am Main, with permission)

Figure 1.6 'Four square' (7-year-old Leon, with permission)

The geographer Jim Blaut was convinced of the deeply imprinted nature of such mapping skills in children, and he understood well, I think, the deeply historical and material origins of such cultural practices (Blaut 1991). In this broadest of senses, he understood what it means to speak of geographical knowledge in a world so thoroughly suffused with carto-graphies of one kind or another, and how cartographic reason has contributed to the ways in which we dwell historically and geographically on the earth (see Blaut 1993). Nearly twenty years earlier Zelinsky (1973: 1) had been blunt in his assessment of the state of our understanding of the significance and reach of this particular form of cultural politics when he argued that 'we have as yet no truly adequate definition of the map; and we have scarcely begun the serious study of its grammar. Thus we do not really know the fundamental nature of the thing we are so intimately enmeshed with, nor do we know what it is we really see or how we think when we look at it.'

Ten years ago I could also write that 'the theory of maps has received comparatively little attention amidst the burgeoning literature dealing with the new theoretically informed geography' (Pickles 1992b). But I think this is no longer true today. While much cartography is still largely concerned with technical issues dealing with the transformation of space and remains committed to representational epistemologies, critical mapping theories and practices have blossomed in recent years.[4]

From the realist ontologies of regional description to the deep anti-representational logics of the 1960s and 1970s that sought to redefine mapping in structural–mathematical or relevant political terms to the spatial turn in the humanities and social sciences in the 1980s and 1990s, mapping metaphors and mapping practices have gained wide currency. The role of maps and mapping in the construction of socio-spatial identities has become an important area of new mapping studies particularly as digital mapping has begun to influence many more domains of social life. In geography, Mark Monmonier (1985) has been insistent about the potential public utility of cartography and prolific in producing a wide-ranging array of books that bring maps and mapping to the attention of a wider public (1989, 1991, 1993, 1995, 1997, 1999, 2001). But it is perhaps in texts from outside of the disciplines of geography and cartography that a new excitement about maps and mapping is most clearly evident. From the wonderful free-ranging interrogation of new mapping technologies in Hall's *Mapping the Next Millennium* (1993) to the rich intertextuality and deconstructive reading of King's *Mapping Reality* (1996) to the call for new cognitive mappings that are found in the work of Fred Jameson, maps and mapping have achieved a renewed status. Geographers and cartographers have been both delighted and shocked to see this flourishing of a cartographical imagination; delighted to see interest directed towards maps and mapping practices, but shocked at both the lack of engagement with professional cartography and cartography's own failure to engage critically with the lives of maps in a broader historical and cultural politics. The result has been a broadening of the mapping impulse in non- and multidisciplinary perspectives, and the proliferation of mapping techniques from territorial mappings to astronomy to mapping the interior spaces of the human body (see Hall 1993). The effects on 'mapping practices' have been electric:

> Listening to veteran practitioners of cartography, one hears laments that yet another elegant academic discipline, refined during the Enlightenment, is being deconstructed, decoded, and destroyed. But the turmoil in cartography signals a great deal more. A craft that once languished in the outback of academe has become a mainstay in innumerable disciplines, all of them seeking to visualize, or map, their data.
>
> (Hitt 1995: 25)

In their introduction to *Mapping the Subject: Geographies of Cultural Transformation*, Steve Pile and Nigel Thrift (1995: 2) suggest that:

> The human subject is difficult to map for numerous reasons. There is the difficulty of mapping something that does not have precise boundaries. There is the difficulty of mapping something that cannot be counted as singular but only as a mass of different and sometimes conflicting subject positions. There is the difficulty of mapping something that is always on the move, culturally, and in fact. There is the difficulty of mapping something that is only partially locatable in time–space. Then, finally, there is the difficulty of deploying the representational metaphor of mapping with its history of subordination to an Enlightenment logic in which everything can be surveyed and pinned down.

In response to these difficulties, Pile and Thrift turn instead to another way of thinking of mapping – as 'wayfinding'. I shall return to this notion of 'wayfinding' later in this book. But, for the moment, I want to invert this question of the difficulty of mapping. Instead of focusing on how we can map the subject, I want to focus on the ways in which mapping and the cartographic gaze have coded subjects and produced identities. My concern has to do with the ways in which mappings function: how they act, in what context, and what are their effects?[5] That is, I am interested in the work maps do, how they act to shape our understanding of the world, and how they code that world. The rapid pace of change in mapping technologies and the forms of visualization and representation that are becoming possible are paralleled by an equally surprising historical continuity in the functions of maps and the institutional and social commitments they support. This book therefore seeks to hold in tension the astonishing technical changes and emerging possibilities for producing new subjects and objects in the world with the continued dominance of state and corporate sponsorship for the development and use of these technologies. At every turn, I try to remain open to the ways in which other producers and users of maps, as well as other mappings that are shaping our understanding of maps, are at work in and between these institutional and corporate spaces. In this sense, the book is part history of mapping sciences and technologies, part social theory of mapping practices and discourses, part critical geography of mapped spaces, and part cultural and science studies of spatial representations. In this sense, the book charts the landscape of mapping as a visual regime – a geography of visual representation – a mapping of the mapping impulse itself. As Tom Conley (1996) has shown, the issue of cartographic metaphors runs to the very heart of western thought itself. It is not only that maps have shaped identities and spaces, but also that the cartographic imagination has influenced the very structure and content of language and thought itself.

A History of Spaces is inspired by a parallel project begun by Brian Harley (1992a: 232), who saw the goal of his own work as showing 'how cartography also belongs to the terrain of the social world in which it is produced'. Harley always stressed the broader contexts within which professional map-making emerged and worked to provide a necessary antidote to the conceit of professionalism and scientism he saw as characteristic of the field. In this sense, the editor of *The New Nature of Maps* (Harley and Laxton 2002), a recent posthumous publication of Harley's texts, refers to a new arena of mapping studies which begins with the understanding that:

> Cartographers manufacture power. They create a spatial panopticon. It is a power embedded in the map text. We can talk about the power of the map just as we already talk of the power of the word or about the book as a force for change. In this sense, maps have politics. It is a power that intersects and is embedded in knowledge. It is universal.

Harley's specific interests were with more traditional map forms, specifically with historical maps. Here, the rational scientific practices of Enlightenment cartography and its commitment to representing 'the real' coincided with the interests of the printer/state interest. The result was an ambiguous form of state cartography producing maps for popular consumption; state cartography democratized access to spatial information, but it did so by prioritizing the interests of the military, the state and private property in its selection of objects to map and the symbolization to be used. This double crisis of representation – democratizing information while representing specific interests – forms the central concern of *A History of Spaces*. How did the cartographic (and corresponding geographical) imagination come to take the form it has? How has the view from space, the God's-eye view, come to typify and so structure our contemporary way of thinking and mapping? In what ways and with what effects have projection as a form of representation, accuracy as a measure of value, and correspondence as a yardstick of truth, come into being? What is this cartographic gaze that so mediates the nature of the geographical in the modern age and seems to exercise so many critical social theorists at the present time? And how, in recent years, have these foundations begun to disintegrate and be replaced by much more plastic and malleable forms?

Wherever possible I have tried to read against the grain of representational epistemologies, what Helen Couclelis (1988) called observer epistemologies and what Richard Rorty (1980) has called the 'mirror of nature'. This has also meant reading more traditional texts and claims without seeing them only in terms of what Donna Haraway (1991: 150) has called the God-trick. I hope in this way to add to David Livingstone's (1992) episodic histories of geographical thought a notion of multiple epistemologies, viewpoints, and visual systems that not only structure the geographical imagination, but are also themselves thoroughly geographical. In so disseminating the history of spaces and knowledges, we are forced to ask a very interesting question: what if, after all, cartography and map-making never really did work with an epistemology and representational economy of reflection?[6]

To the long and distinguished story of maps in the history of cartography and geography, I have also turned outside these disciplines. The relationship (and the tensions) between these two groups of mappers and writers is a useful point of departure for the present work. It is in this nexus of professionalized and popular map-making and map-readings that a series of themes that run throughout the book can be identified. Too often the story of maps (in both the popularized and professional literatures) has been captivated by the history of scientific advancement and individual achievement; too focused on technical progress and the progress of 'accurate representation'. By this I mean that the story of maps has been, and remains, for the most part a story of technical achievement and the advancing capacity of cartography to represent the earth and its geographies, mixed with a general incredulity towards issues of metaphor, symbol and myth. Non-representational mappings of indigenous and non-modern peoples (who are generally equated with one another) have fared poorly in this treatment. In this view, it is in the modern form that the 'representation of the real' reaches its zenith and places cartography and the map at the heart of a scientific Enlightenment project. By contrast, traditional forms of mapping (such as 'religious' T-and-O maps to Pacific islanders navigational 'stick' charts to medieval triptychs) are explained in terms of mythic iconography and approximate knowledge (Figure 1.7). With science and the Enlightenment, these 'strange' lands and their maps become the terrain of rational calculation and the representation of the real: accuracy, correspondence and detail become the hallmark of nineteenth-century mapping projects and the basis for the stories cartographers tell today.

As a result, this scientistic reading of the story of maps and the craft origins of the field of cartography have each inscribed in our understanding of maps a narrow field of what counts as a legitimate map. In this book I hope to show how maps and mapping can be thought in much broader terms and in ways that enable us to open the contemporary meanings of the map for social inquiry. Consequently I develop a catholic reading of maps and mapping that includes, along with the printed maps of the craft cartographer and the digital maps of the computer cartographer, schematic representations of a wide variety. Others have similarly recognized the need to open the canon of maps and mapping to a wider range of forms and practices. For example, Denis Wood's 'cartography of reality'

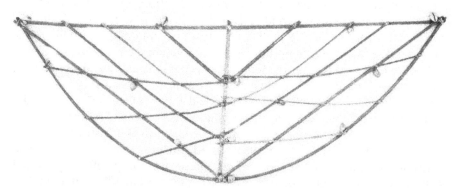

Figure 1.7 Marshal Islands stick charts are constructed by palm ribs bound by coconut fibre with shells used to represent islands. The stick charts function as instructional and memory devices to understand swell patterns and island location. There are three main kinds of stick chart: *mattang*, *meddo* and *rebbilib*. The first and second use the palm ribs to indicate swell lines, with some use of shells to represent islands. The latter provide more detailed information about islands and island groupings. Charts are constructed by individual navigators for particular purposes and are, as a result, not thought of as generalizable (Turnbull 1993, *Maps are Territories*, with permission, University of Chicago Press)

called for explicit recognition of the importance of the ways in which ordinary people map their worlds and use their maps in map-making societies, Roger Downs suggested that the 'maps in our minds' play an important role in social action, and Bill Bunge urged us to use our map-making skills to produce maps for a people's geography.

This issue is the explicit focus of several volumes of the *History of Cartography*. In Volume 2(3) entitled *Cartography in the Traditional African, American, Arctic, Australian and Pacific Societies*, Woodward and Lewis (1998: 2) made clear that earlier definitions of 'map' and 'cartography' had been too western in orientation and too limited in scope: 'By using the word "map" to cover so many different things, we are simply extending the logic of earlier volumes that called the Greek *pinax*, the Roman *forma*, the Chinese *tu*, and the medieval *mappa mundi* and *cartada navigare* "maps" and included them in a cartographic history.' In dealing with such indigenous non-western cartographies, it has also been necessary to rethink definitions of the map. Indigenous mappings do not necessarily have the same kinds of materiality and reproducibility as do western maps, and what constitutes a map and a mapping practice is not necessarily the same across cultures. In some societies, gestural and performative practices are central to the ways in which people structure and represent their worlds spatially, serving as tools of way-winding and spatial representation. Even where no formal cartography emerged prior to European contact, mapping (with all of its performative and material implications) certainly did. Thus, in the latest volume in the *History of Cartography*, 'mapping' instead of 'map-making' has been used to determine what counts as a map, greatly expanding the domain of 'maps' to include a wide variety of representational and symbolic forms under the rubric of the 'history of cartography' (Table 1.1). In this sense, pre-modern cartography and indigenous cartography now have such a 'history', one which seems almost certain to revise the genealogy of modern cartography itself.

This recent reopening of the cartographic canon to the cognitive, performative, semantic and symbolic richness of mappings, as well as the diversity of material products that embody those mappings, even more sharply highlights the limits of traditional cartographic thought. In scientific cartography, narrower definition of maps as mirrors of nature have tended to define many analytical, metaphorical and symbolic representational systems out of the cartographic canon. For example, I have long been intrigued by the deft way in which spatial models of economic surfaces have not entered into the lexicon of cartography and play very little role in the history of cartography. Von Thunen circles and central place hexagons – perhaps two of the most iconic cartographic forms in twentieth-century spatial science – play virtually no role in the story of maps. And yet what fascinating cartographies these abstract analytics produced and what better examples can there be of the coding of landscapes and social action than, for example, Brian Berry's (1966) integration of

Table 1.1 Categories of representations of non-western spatial thought and expression (courtesy David Woodward)[7]

Processes		Products
Thought	**Performance**	**Record**
Cognitive cartography	**Performance cartography**	**Material cartography**
Organized images such as spatial constructs	*Non-material and ephemeral*	*In situ*
	Gesture	Rock art
	Ritual	Displayed maps
	Song	
	Poem	*Mobile comparable objects*
	Dance	Paintings
	Speech	Drawings
		Sketches
		Models
	Material and ephemeral	Textiles
	Model	Ceramics
	Sketch	Recording of 'performance maps'

maps and graphs page after page to produce a landscape of spatialized commercial transactions in India.[8] Or what epistemological shifts in mapping theory were at work in the mathematico-structural mappings of Peter Gould's interest in multidimensional space (Figure 1.8)?

In thinking about the social lives of maps, I want to introduce these kinds of mappings (from experiential to cognitive to popular to spatial analytic) into the story of cartography to see how their presence changes the story we tell. But in introducing them I want to do more than merely enlarge the canon of what counts as a legitimate and interesting map.

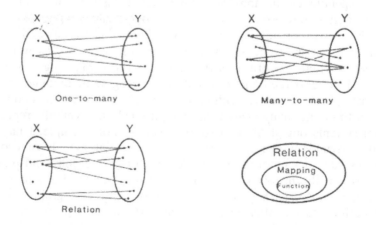

Figure 1.8 'Relation ... Mapping ... Function', Peter Gould (with permission, Jo Gould)

By so juxtaposing, for example, spatial analytic models to craft maps I want to suggest that (at least) two mapping regimes, two discourses, institutions, and sets of practices are at work in the mapping process that sit uneasily next to each other. One has its roots in the craft guilds of early capitalism. The other has its origins in attempts to map the emergence and workings of late capitalism. These distinct (though overlapping) political economies are articulated in important ways in the story of maps, and provide us with the second reason for juxtaposing the two. Thinking this juxtaposition as one rife with contradictions and tensions also allows us to think about cartography and cartographic reason in new ways. As Gregory (1994: 7) has warned us, critics must be careful not to 'accept cartography's own historiography even as they contest it'. 'It is', he goes on:

> perfectly true that historians have usually presented cartography as the Survey of Reason, a narrative journey of progress from darkness to enlightenment, in the course of which maps become supposedly more 'accurate' and more 'objective.' But it is also true that there is now a critical historiography, which has established the implications of maps in the constitution of systems of power-knowledge and, through the works of Brian Harley in particular, has suggested ways of deconstructing their technologies of power.
>
> (1994: 7)

In this sense I want to evoke Gunnar Olsson's suggestion that in the dash and the line can be found the history of cartography. But, in accepting this suggestion I do not also accept that the 'notion of map-able space involves a specific epistemic topography; a landscape, a form of knowing or seeing which denies its structuring by the gaze of white male bourgeois knowers on Other knowns' (Pile and Rose 1992: 131). By thinking through 'power-talk' in the critical historiography of mapping I try to show how, far 'from limiting the possibility of critique by refusing to acknowledge other kinds of space' (Pile and Rose 1992: 131), cartographic reason and the project of mapping have been contested and multiple from the beginning.

As he challenges cartography to become conceptually and analytically more rigorous, Olsson offers a way of thinking about mapping that generalizes mapping and its products, and thereby denaturalizes and deprofessionalizes cartography. In this way, he locates the epistemology and technics of representation that constitute the field of cartography in the context of social practices beyond the profession of cartography. This seems almost a commonplace today as we look at the proliferation of mappings in geographic and demographic information systems, in geographical positioning system (GPS) and surveying, in popular advertising, and games for children and adults. But it is not enough to think of these mappings as mere variants of the cartographic imagination. They extend our thinking and our practices into domains that might not initially be thought of as cartographic, but that have been and can be again important moments for thinking about how we represent the spaces and places of our world, and how we map. One can think, for example, of the eighteenth- and nineteenth-century fascination with exhibition and model (both important technics in the emerging political economy of western capitalism and the liberal state) and the ways in which representational technologies of panorama, diorama and museum paralleled and influenced the emergence of modern cartography (see Figures 5.5, 7.1 and 7.5). In a similar way our own tele-cultural fascination with photographic, cinematic, and televisual images (important technics in the political economy of high mass consumption) are clearly important in the ways in which digital mapping has articulated its own representational power. As David Harvey (2000) has reminded us, we need to be much more attentive to the institutions within which geographical and cartographical knowledges are produced and disseminated. Harvey identifies the usual institutional suspects (state apparatuses, military power, supranational institutions, non-governmental organizations, corporate and commercial interests, media and tourism, and education and research institutions), each producing distinct, overlapping and particular forms of governmentality.

Such 'productions' and 'disseminations' are always productions and disseminations of dialectical images. The delimitation of territories and identities through the dash and the line is at one and the same time a bounding and separating that does violence to the world *and* a practice that gives our present world the meaning we understand and use on a daily basis. The categories we use and the demarcations we draw produce identity/difference relations in terms of which the world is structured and understood. This is what we mean when we say that the world is socially constituted or produced. It is not a matter of personal choice, but a matter of pre-existing and changing forms of identification, categorization and typification that transform and rework structures of meaning and identity and their corresponding spaces. Spaces are always being – and can be again – reconfigured by new geo-graphs and carto-graphs, new writings, new lines of inscription and new lines of demarcation. Recognizing the socially constituted nature of identity claims (our concepts, categories and practices) is a first step to a deconstructive retrieval of other possible worlds, spaces and mappings.

No longer is the cartographic imagination bound by the territorializing impulse of the nation-state, the imperial project, or commercial and scientific exploration. The globalization of capital, information and culture, the internationalization of economic and political practices, and transborder, transnational and diasporic migrations have all necessitated a parallel deployment of new analytical categories and new spatial descriptors. In turn, they have contributed to the flourishing of renewed geographical, geopolitical and cartographic imaginations. For David Harvey (2000) 'mapping space' is 'a fundamental prerequisite to the structuring of any kind of knowledge. All talk about 'situatedness', 'location' and 'positionality' is meaningless without a mapping of the space in which those situations, locations, and positions occur. And this is equally true no matter whether the space being mapped is metaphorical or real' (Harvey 2000: 111–12).

What then does it mean to map space and to make and use maps? How are maps drawn and from where does their power derive? From what position do we view the world in order to construct the maps we draw and use? In pulling apart distinctive scopic regimes (such as perspectivalism and projectionism) each based on their own geography and political economy, Martin Jay has demonstrated what geographers Alan Pred and Michael Watts (1992) earlier characterized as the emergence of multiple modernities. It seems to me that this multiplicity, and the differences it fosters, are at the heart of the problem of coming to grips with so ambiguous and complex a cultural object as the map. The still deeply rooted desire for totalizing monochromatic accounts that explain the map in terms of it being a socially produced symbolic object, a tool of power, a form derived from a particular epistemology of the gaze, or a masculinist representation, seem to me to miss the point of the post-structuralist turn: that is, that not only are maps multivocal, not only are the spaces they constructively represent complex articulations of coded and nomadic spaces, but so also must be our accounts of them. I think here with Deleuze and Guattari (1987) who argued against the binary logics of either/or and in favour of the multiplying, rhizomatic logics of 'and ... and ... and...' that multiplies interpretations and readings, providing more tools, more languages, and more assemblages to help us in thinking about the various kinds of work maps do. As Matthew Sparke (1998: 464) has argued: 'while scholars such as Benedict Anderson have discussed the general *hegemonic* effects of national mapping, they have rarely addressed the *counterhegemonic* effect of cartographic negotiations'. Through what he calls 'the supplementary performances' of map-making and map use, the homogeneity of hegemonic projects and readings is transformed into unavoidable heterogeneity (p. 147) resulting in what he calls 'contrapuntal cartographies'.[9]

MAPPING THE BOOK

A History of Spaces aims to contribute to the emerging critical literature on the nature of maps and their use in the contemporary world. This book is neither a guide to map-making and map use, nor a history of cartography. It is, instead, a mapping of cartographic practices, institutions and discourses – a genealogy of the map and its social roles – which locates the map within broad historical, social and political contexts.

The book draws on contemporary social and geographical theory, and contextualizes mapping practice and map use within a critical analytics of science, technology and society. Specifically, this involves three critical moments that run through all chapters of the book. The first has to do with *dreams*: with the wish images of the mapping enterprise, of cartography and of what Derek Gregory has called 'the geographical imagination'. The second has to do with *magic*: with the ways in which science, expertise and accuracy have been deployed to transmute the world into tradeable values. Representational economies of mapping, map-making and map use have literally coded the modern world in ways that have turned geography into gold. In this second moment I focus on how particular understandings of vision and representation (a particular scopic regime) have been co-present within the internal and public narratives of mapping, and how these must be understood as central to any consideration of the power–function of maps. The third moment has to do with *performance*; with the ways in which maps and mapping function in contemporary society, and how they produce subjects and constitute identities. Thus, the book aims in an indirect way to answer three questions: (a) how do maps produce subjects? I shall call this investing subjects in depth; (b) How do maps produce bodies? I shall call this investing bodies in depth; and (c) what are the implications of the current fascination with three-dimensional virtual realities, imaging systems and simulations? I shall call this investing space in depth.

Part II begins with a deconstruction of the map through a reading of the works of Brian Harley, showing how the everyday notion of the map and map use permeate scholarly and professional discourses, and how – as Harley argued – these 'disseminated' relations of power must be questioned. At one level, the map and the mapping exercise can be seen as the careful scaling and coding of worldly objects and spaces for particular purposes: the topographic map enables accurate assay of and navigation through the landscape; the geological map identifies regions of similar surface and subsurface rock, along with boundary features such as faults and fracture zones; the architectural plan identifies the inner and outer spaces of built objects to guide the builder, lawyer and owner; and the street map identifies property boundaries, public infrastructure and official names for buildings, streets, and public and some private spaces. At another level, the map has emerged as a tool (or technology) embedded in a set of practices and institutions that affect the ways in which we live our lives in the modern world – a way of *cataloguing* the 'important' (and ignoring the 'unimportant') features of the earth's surface and the social world; a way of *accounting* for the resources, objects and public infrastructure of the earth's surface; and a tool for the *representation* and *territorialization* of space. That is, by tracing genealogies of mapping, I ask how the map emerged as a tool of a science wedded to representational thought and observer epistemologies; how the mapping impulse emerged and changed from pre-modern to modern forms; how the map serves as both scientific tool and cultural icon; and how the map has served various roles within the experience of modernity. This is followed by a discussion of propaganda maps and the underlying theory of maps that contemporary map use presupposes. In particular, this section elaborates the roles played by the map and map metaphors within earth and social sciences. Chapter 3 focuses on map use. I build on Denis Wood's (1992) analysis *The Power of Maps*, but focus more explicitly on the map as a practice and discourse that names the world, categorizes people, bounds places, and territorializes socio-politico-economic regions.

Part III comprises four chapters dealing with some of the concrete ways in which maps code worlds. These chapters focus on the visual geographies of modern mapping, cadastres and capitalism, the geo-body of the territorialized state, and the categorization and commodification of social life. Chapter 6 takes up again the theme of the power of maps and questions the 'power talk' that has come to dominate social constructionist readings of maps. In deconstructing 'power talk', I draw on a variety of sources and perspectives to decentre such ideas, and instead focus more on their geographies, conceptual leakiness, and their roles in the transculturation of ideas. Chapter 7 turns to a cultural politics of mapping and focuses on the mapping of modern subjects as a constitutive process. Turning away from repressive understandings of power, this chapter seeks to articulate a notion of maps in terms of productive power – as constitutive of the very being of modern subjects. In this way, the book represents a crucial supplement to Harley's power of maps and deconstructing the map. Central to Foucault's genealogical approach to knowledge–power is his insistence on the non-reducibility of the scientific disciplines to general or universal principles of reason, objectivity and truth, and the historically constitutive role of the scientific disciplines in producing the categories and practices we think of as normal or natural (such as mentally ill, abnormal, criminal, citizen, population, etc.). Genealogy is an attempt to trace the relations among specific disciplining complexes of institutions, actors, discourses and practices, and to describe how modern power flows through them. In this way, mapping as power–knowledge produces the subjects we are in varied assemblages of institutional setting (docile subjects, healthy bodies, good citizens, reading publics, patriotic nationals, active individualized consumers, etc.). The chapter thus unpacks more explicitly what it means to think of maps and mapping in anti-essentialist and non-essentialist terms as discourses, practices and institutions within which subjects and identities are produced (with all of their limits and *potentialities*).

Part IV deals with the ways in which new visual imaginaries and new spatializations of objectness are being enabled through digital information and mapping systems. In particular, the chapters in this section deal with what I call 'mapping subjects in depth'. Such 'mappings in-depth' are increasingly possible as information systems and imaging capabilities give rise to new cartographies; transparent depth ontologies, new forms of action at a distance (be it in medical imaging or military imaging technologies), and the ability to interpolate new identifications and typifications from existing databases. The chapter asks how these emerging visual regimes shape new possibilities for and practices of civil society, democracy and social action.

Part V concludes with an investigation of alternative claims for maps and their usage, including Bunge's work on people's maps, debates about gendered spaces, notions of insurgent mapping, and attempts to think of non-representational, denaturalized mapping systems (for example, through the montage of Walter Benjamin, the mapping experiments of Dadaism, the psychogeographies of the situationists, and the spatial thought of Paul Virilio). It concludes by asking whether and in what ways our efforts to reconstruct critical histories of maps and mapping might (must) be understood in terms of broader processes of globalization.

Michel Foucault has reminded us that a whole history of spaces is still to be written.[10] An important part of these histories of spaces is the history of mapping practices and the cartographic imagination, through the map, but also through landscape painting, the panorama, the diorama, the great national exhibitions, the museum, the history of epidemiology, public health, the police, and property development, through the aerial photograph, the satellite and through geographic information systems. This book seeks to contribute a brief history of the technologies and practices of cartography to these broader and richer traditions of mapping and map theory. All the chapters attempt to locate maps and mapping in terms of the practices, discourses and institutions of their production and use. Each tries to hold together (albeit in uneasy tension) the technical changes in mapping practice, the broader technological fields within which those practices emerge, and the material circumstances that call these practices and technologies into being (for example, the demands of commercial exploration in the seventeenth and eighteenth centuries, territorial expansion in the eighteenth and nineteenth century, urbanization and population change in the late nineteenth century, the control and management of Empire in the nineteenth and twentieth centuries, and war economies and societies of surveillance in the twentieth century).

My own concern is threefold. First, to understand how maps and mapping have shaped our world. Second, to see mapping, maps and cartographic reason as central to the geographical imagination in the sense that they are crucial elements of social inscription that produce spatial identities. And third, I want to illustrate the ways in which this graphism that lies at the heart of earth writing (cartography in all its forms) is also a form of thought and practice that permeates more widely – in non-disciplinary channels – in the history of our world. For Olsson (1998: 146), western philosophy itself was cartographic, at least in the ways in which philosophers were 'scouts in the unknown territories of the taken-for-granted, mappers of the boundary between Oecumene and Anoecumene, spies in the no-man's-land between the sensibility of the body and the intelligibility of understanding.' If, as Franco Farinelli (1998: 135) has suggested, the grapheme lies at the heart of logos, it would mean that 'western thought (reason) is nothing else than the protocol of geographical presentation, that is of the cartographic image. Further, this would imply that our rationality is determined from a cartographical point of view, that it is already contained and produced by the cartographic image. Western thought is nothing but cartographical reason...'

But, while Farinelli sees these as determinate of certain forms of rationality, I want to investigate in what ways this cartographic imaginary proliferates spaces and the ways in which we can live in them. That is, it is the very structure of cartographic reason that – far from inscribing a single determinate line – draws and redraws our world, erases and inscribes again, decodes and recodes, in a ceaseless and complex array of forms of deterritorialization and reterritorialization producing the multiple and shifting identities (or assemblages) we take as ourselves. Henri Lefebvre (1991: 85) asked: 'How many maps, in the descriptive or geographical sense, might be needed to deal exhaustively with a given space, to code and decode all its meanings and contents?' His answer is important for any history of spaces and mappings:

> It is doubtful whether a finite number can ever be given in answer to this sort of question. What we are most likely confronted with here is a sort of instant infinity, a situation reminiscent of a Mondrian painting. It is not only the codes – the map's legend, the conventional signs of map-making and map-reading – that are liable to change, but also the objects represented, the lens through which they are viewed, and the scale used. The idea that a small number of maps or even a single (and singular) map might be sufficient can only apply in a specialized area of study whose own self-affirmation depends on isolation from its context ... We are confronted not by one social space but by many – indeed, by an unlimited multiplicity or unaccountable set of social spaces...
>
> (Henri Lefebvre 1991: 85)

Part II

Deconstructing the map

The classical age discovered the body as object and target of power.

(Foucault, *Discipline and Punish*)

Disciplinary space tends to be divided into as many sections as there are bodies or elements to be distributed.

(Foucault, *Discipline and Punish*)

2   What do maps represent?

The crisis of representation and the critique of cartographic reason

It is comparatively easy to visualize maps as representational models of the real world, but it is important to realize that they are also conceptual models containing the essence of some generalization about reality. In that role, maps are useful analytical tools which help investigators to see the real world in a new light, or even to allow them an entirely new view of reality.

(Board, 'Maps as models')

A map seems the type of conceptual object, yet the interesting thing is the grotesquely token foot it keeps in the world of the physical, having the unreality without the far-fetched appropriateness of the edibles in Communion, being a picture to the degree that the sacrament is a meal. For a feeling of thorough transcendence such unobvious relations between the model and the representation seem essential, and the flimsy connection between acres of soil and their image on the map makes reading one an erudite act.

(Harbison, *Eccentric Spaces*)

Maps and mapping have always been of theoretical and practical importance to geographers and cartographers, and they have had long-standing technical and metaphorical importance for the theory and practice of fields such as geology, surveying, astronomy, anthropology, art history and literature. In recent years, the emergence of new capacities of digital mapping in cartography and the spatial turn in the humanities and social sciences have extended mapping practices and metaphors across the social field. As Board foresaw in the quotation with which this chapter begins, a new analytics and a new view of modelling reality have been in the making; a deepening and extension of the possibilities of spatial representation requiring the specific reading skills and erudition to which Harbison points. At the same time, profound epistemological changes have shaken the self-understanding of the sciences and humanities as Cartesian dualisms and scientistic naturalisms of all kinds have been brought into question. The result has been nothing short of a 'crisis of representation' and it is this crisis that forms the core of this chapter. I focus on four main aspects of the broader impact of the crisis and of the spatial turn in social theory:

1   The renewal of the cartographic imagination spawned by the spatial turn in social thought is having an important influence on a broad array of social science and humanistic disciplines and social practices, providing new metaphors and frameworks for thought and action.
2   Cultural studies are being transformed by a rethinking of the carto-graphic imagination in ways that pose challenges to and opportunities for rethinking cartographic practices themselves.
3   Technical changes are blurring any former boundaries between car-tography, imaging, and social and scientific practices. As a result, the theory of cartographic representation that held sway for so many years has begun to show signs of wear and tear. In particular, new technologies and uses of spatial representations have brought to the forefront again issues of accuracy and error.
4   The broadening and deepening of mapping practices reflected in the first of these three points presents a paradox for geography and car-tography. In recent years, we have become much more aware of the many ways in which cartographic reason has underwritten the struc-ture of thought in other fields. But the recognition that its own imagi-native structures are still firmly rooted in representational logics and beliefs has come late to geography and especially cartography. What Gregory called the 'Cartographic Anxiety' and the epistemology of viewer and world, subject and object, interiority and exteriority on which it rests continues to limit the theoretical and practical possi-bilities of cartography itself.

In the next section, I unpack the Cartographic–Cartesian Anxiety in terms of three related elements of the crisis of representation. The first focuses on the emergence and role of communication models of informa-tion and the objectivism to which they laid claim. The second turns to J.K. Wright's arguments about the subjective nature of maps. The third deals with how implicit assumptions about objectivism and subjectivism frame the understanding of error and distortion in cartography. I elaborate these three elements not to provide a thorough synthetic genealogy of carto-graphy's self-image or self-understanding, but to highlight one aspect of this self-understanding, its abiding Cartesianism and the depth and consequences of this particular commitment.

THE SPATIAL TURN IN SOCIAL THEORY AND NEW
SOCIAL AND CULTURAL CARTOGRAPHIES

> In terms of most communication theories and common sense, a map is a
> scientific abstraction of reality. A map merely represents something which
> already exists objectively 'there'. In the history I have described, this rela-
> tionship was reversed. A map anticipated spatial reality, not vice versa. In
> other words, a map was a model for, rather than a model of, what it pur-
> ported to represent ... It had become a real instrument to concretize pro-
> jections on the earth's surface. A map was now necessary for the new
> administrative mechanisms and for the troops to back up their claims ...
> The discourse of mapping was the paradigm which both administrative and
> military operations worked within and served.
>
> (Thongchai, *Siam Mapped: A History of the Geo-Body of a Nation*)

The underlying changes that have brought about a crisis of representation
and a re-engagement with cartographic reason are particularly well illustrated in Stephen Hall's (1993) *Mapping the Next Millennium: How
Computer-Driven Cartography is Revolutionizing the Face of Science.*
Perhaps the landmark of this contemporary broadening and resituating of
the mapping impulse in non- and multidisciplinary perspectives, *Mapping the
Next Millennium* illustrates particularly clearly the proliferation of mapping
techniques and uses ranging from territorial mappings to astronomy to
mapping the interior of the human body. In their sheer scope of coverage,
such mapping systems and practices challenge the reader to ask how these
different 'ways of seeing' have roots in particular regional political
economies, cultural geographies and historical traditions, and how such multiple, but parallel and linked, forms have been grafted onto, and still influence, contemporary representational practices (see Cosgrove 1988, 1989).

   Mapping the Next Millennium is part of a broader canon of works
dealing with the changing nature of mapping practices. These have
encouraged much more attention to the ways in which maps inscribe and
shape socio-spatial identities (for example, the naturalizing of new social,
class and neighbourhood categories in Charles Booth's maps of London or
in the Hull-House maps, national identity in Thongchai's *Siam Mapped*, or
self and state in Renaissance France in Conley's *The Self-Made Map*).[1]
They have expanded the ways in which the cartographic impulse is understood in ways that go well beyond traditional disciplinary frameworks.

   Rolland Paulston's (1996) *Social Cartography: Mapping Ways of Seeing
Social and Educational Change* is an interesting example of this new post-
disciplinary mapping that draws explicitly on a wide range of contemporary social theorists, social scientists and humanists. For Paulston, mapping
– as a fundamentally non-linear representational system – provides
a means not only for rendering concrete representations of patterns,
but also for opening up the spaces of *thinking* and *discussion*. After

encountering the spatial turn in the work of geographers while visiting at the University of British Columbia and working through new writings in phenomenology, postmodern geographies, works by Bourdieu, French post-structuralists and feminist cartographers, Paulston (1996: xvi–xvii) says he was better able to 'understand how the spatial turn in comparative studies would focus less on a formal theory and competing truth claims and more on how contingent knowledge may be seen as embodied, locally constructed, and re-presented as oppositional yet complementary positionings in shifting fields.' By drawing on Henri Lefebvre's resistance to categorization and Gilles Deleuze and Felix Guattari's call for 'nomad' mapping, *Social Cartography* aims to think beyond the binary categories of what Paulston calls an 'intentionally modernist mapping of social cartography'. The intent is 'the crafting of a ground-level social cartography project with critical potential, that would build upon and extend earlier postmodern mapping contributions in geography and also in feminist, literary and postcolonialist studies. Work in this new genre uses spatial tropes to map discursive fields', it rejects essentialism and scientism, it understands contemporary knowledge as 'akin to a space of shifting sites and boundaries definable only in relational terms' (Paulston 1996: xvii), and it accepts Soja and Hooper's suggestions that 'this spatialized discourse on simultaneously real and imagined geographies is an important part of a provocative and distinctly postmodern reconceptualization of spatiality that connects the social production of space to the cultural politics of difference in new and imaginative ways' (Paulston 1996: xvii quoting Soja and Hooper 1993: 184).

Social cartography is 'the art and science of mapping ways of seeing' that seeks to avoid the rigidities of traditional mapping practice by shifting the focus to the efforts of individuals and cultural groups to define their own 'sociospatial relations and how they are represented' (Paulston 1996: xv and xviii; see also Paulston 1997: 117–52). Social cartography is thus a mapping of relational spaces orientated 'toward charting the variable topography of social space and spatial practices', understanding how 'sliding identities' are created, and finding ways to represent these motions in ways that reflect their contested and discursive origins (Paulston 1996: xviii–xix).[2]

Geoff King's (1996) *Mapping Reality: An Exploration of Cultural Cartographies* is similarly concerned with the crisis of representation in modernist thought. In replacing the objectivism of representational thought with a discursive analysis of the processes of mapping and identity formation, *Mapping Reality* is a timely intervention in the 'power of maps' and 'maps as power' literature. In these readings, what Foucault (1984) called repressive notions of power hold sway and interpretations of maps and mapping that reduce the map to this or that embodied interest have proliferated.[3] Early empiricist readings of maps (where maps were seen to be the unproblematic representations of an external reality) have thus increasingly been replaced by reductionist readings of the power of and in maps. These have been productive in the ways in which they have challenged empiricist and technicist readings of maps, but limiting in their tendency to reduce theories of mapping to theories of power. In the place of this repressive notion of power King provides a more culturally situated and conceptually nuanced reading of maps and mapping as concrete historical practices. The result is a remarkable tour de force; an ambitious reading of mapping practices and map uses through three centuries, a rich selection of themes and detailed case studies, and a critical deconstruction of maps and the mapping enterprise.

Map and territory, image and reality, as binary constructions of a modernist world do not survive long in King's text. The first page of the book begins with a disconcertingly straightforward reading of Garrison Keillor's mythical Lake Wobegone which does not appear on the map because '[m]istakes were made by cartographers working without the benefit of aerial views or modern technology' (King 1996: 1). It ends with Jean Baudrillard's claim that the map has come to precede the territory. Instead of thinking of the map as the product of a territory or a passive representation of it, King (with Baudrillard) suggests a strategic reversal; it is the map that engenders territory. The notion of the real as something existing in its own right is no longer tenable. The real is not only what can be reproduced, but that which is already reproduced – the hyperreal. Thus

> Map and territory cannot ultimately be separated. Cultural mappings play a central role in establishing the territories we inhabit and experience as real ... To blur this distinction between map and territory is to destabilize this relationship, to acknowledge the socially constructed character of the mappings within which our lives are orientated.
>
> (King 1996: 16–17)

Particularly interesting in *Mapping Reality* is the way in which the author evokes complex readings of maps through story after story about their development, use and effects. The text conjoins a refreshing theoretical richness with an equally refreshing level of concrete detail about the map in various contexts. King offers us a reading of maps and mapping that is multiform and complex, without reducing one form or interpretation to another. Throughout the book, King raises questions of power, interest and alternative uses and rejects standard cartographic approaches which understand mapping in terms of information theory, or that make claims to authority based on the neutrality, objectivity or *transparency* of maps.[4] Instead, maps are to be understood as products of particular representational practices. 'World-views' are the material products of cultural projects such as nation-building, colonial expansion or cultural hegemony. In each, maps (and other forms of representation) have played their role. Systems of meaning are inscribed in maps through the lines, boundaries and symbols that give meaning and reality to the world. These are not mere representations of reality but come to represent objects whose existence is in part conditioned and produced by their representations.[5]

Mapping Reality probably fails in one important sense for geographers and cartographers in that it does not undertake any systematic engagement with recent geographical work on maps and mapping. While Brian Harley is present to a limited degree, even standard geographical texts such as Norman Thrower's (1972, 1996) *Maps and Civilization: Cartography in Culture and Society*, Denis Wood's (1992) *The Power of Maps*, or David Turnbull's (1993) *Maps are Territories* are not considered. Even *The History of Cartography* project does not make it into *Mapping Reality*, and a myriad of critical hermeneutic, Marxist and post-structuralist readings of maps and mapping are also absent (e.g., Harley and Woodward (1992), Pickles (1992b), St. Martin (1995), and the various essays on cartography and mapping in Reichert (1996)).

This important lacuna is, surely, indicative of something more serious than disciplinary focus and trans-disciplinary oversight. I think of it as a paradox that might help us to understand one of the central problems of geographical and cartographical imaginations at work. While Hall opens our eyes to the development and diffusion of new cartographic and imaging technologies and practices, and Paulston builds on his encounter with geographical texts and ideas but in the end remains committed to a modernist project of synthesis and totalizing mapping, King has cut loose the foundational tethers and allowed his readings to focus on multiple discursive formations. But he has done so at the expense of any engagement with the work of contemporary geography and cartography. While one could ask of these three authors to be more attentive to the literature of geography and cartography, I prefer to turn around the problem of the lacuna and ask, instead, what is wrong with contemporary cartographic theory and practice that this can happen at a time of such growth in the mapping sciences? My short answer is what I will call the paradox of representation and its commitment to objectivist epistemologies of science, or what Derek Gregory has called 'the Cartographic Anxiety'.[6] This paradox brings us to the three elements of the crisis of representation.[7]

FIRST CRISIS OF REPRESENTATION: THE OBJECTIVISM OF SCIENTIFIC CARTOGRAPHY

For geographers and cartographers, 'The notion of a "map" ... is essentially that of a model, a representation of a geographical area (usually) on a flat surface. Ordinarily, each point on the cartographic diagram corresponds to an actual geographical position on earth, according to a definite scale or system of projection' (Henricksen 1994: 52). Maps 'serve as the base to register geographic data', by facilitating the inspection of distributional patterns they help the researcher to 'uncover possible relationships', and they serve 'to communicate the results of research in more generalized form' (Jan Broek 1965: 64).

By the 1960s and 1970s, this understanding of cartography as representation and communication was increasingly being articulated in terms of 'communication science' in which the map functioned as a tool for communicating spatial information (Robinson and Petchenik 1976).[8] In this view, maps were devices of information transmission involving the basic rules of communication (source-channel-recipient) based on a one-to-one correspondence of the world and the message sent and received (Muehrcke 1972). At the time, such communication models of information were common across the social sciences. These had been greatly influenced by the adoption of informational models of the mind in psychology, with scholars such as Robinson (1952) suggesting that a new cartography might be grounded in experimental psychology. Such information models were also stimulated by research at techno-scientific research labs such as those at Harvard, MIT, Berkeley and Bell Labs, and those more directly funded by the US Government's combined efforts to both build Cold War security institutions (e.g., Ciccone *et al.* 1978, Martin and Rinalducci 1983) and rebuild the cities (e.g., Craik 1977). Muehrcke's model of the cartographic processing system drew on these wider debates about information and communication, but in practice cartographers soon settled on a rather more instrumental approach to map use and on models that assumed a more mechanical transfer of information from the 'real world' to 'raw data' to the 'map made from raw data' to the 'user's mental image of the map' (Kimerling 1989: 688). In the process, the complex processes of meaning and metaphor were gradually being lost as process models understood in terms of sender (inputs), medium (transfer) and receiver (outputs) models of communication held sway (Monmonier 1975).

This model of communication required that information from the sender be encoded and that the receiver decode the information. Information is conveyed, and, in so far as the cartographer, map, and map-reader all receive the same information, distortion is avoided (Robinson and Petchenik 1976). The measure of communication efficiency in the mapping process is related to the amount and accuracy of information transmitted. The cartographer's task is to devise better approximations between raw data and the map image (Muehrcke 1972) and the map-reader's responsibility is to interpret the symbolization of the map carefully and accurately. The map itself is merely an objective tool for transmitting this information. In so far as the technical production does not distort the data collected from the 'real world' the 'good cartographer' is successful, and in so far as the map-reader interprets the information accurately he or she is a good map-reader. In both cases, the primary responsibility in handling maps is to manage error technically and with skill. This is a form of realism and representationalism that is anything but 'naïve'.

When asked about the representational nature of maps any cartographer will point out that maps are always a compromise among error terms. Of all mapping techniques, map projections most clearly illustrate this. Flattening the surface of the globe inevitably produces distortions. More technically, mapping a two-dimensional surface of constant positive curvature on to a planar surface involves transformation of some combination of shape, area or directionality (azimuth). Hence, as the old saying warns, 'All maps lie flat, therefore all maps lie' (Henricksen 1994: 52).

Map-makers have always understood the importance of choice in map design; not only is the world too full to represent everything, but sometimes important information is not available. As a result, the art of map-making has been tied closely to efforts to formalize and sharpen the nature of the transformations involved in projection (Figure 2.1). Once an appropriate projection has been selected to achieve minimum distortion in terms of specific criteria (area, shape or azimuth), the map-maker's task continues to be one fraught with difficult choices and interpretations, inclusions and exclusions, thicknesses and thinnesses, additions and erasures. How does the cartographic imagination render geographical patterns in map form? How are lines chosen and how are they measured, drawn and circulated? And how does it happen that even simple line drawings come to mean so much in the practice of worldly affairs? How do maps work so well? Consistent selectivity has been the hallmark of all cartographers and

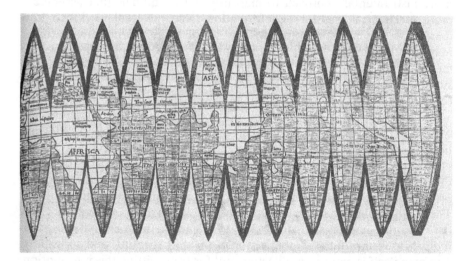

Figure 2.1 Waldseemüller's terrestrial globe gores, 1507. Cartographers have long been familiar with the technical challenges of dealing with error and distortion resulting from all mappings of a two-dimensional surface of constant positive curvature, such as the earth's surface, onto a two-dimensional planar surface. Much less attention has been paid to the social and moral conditions and consequences of such 'renderings' of the earth through gores, slices and projections

cartographic institutions (Wood 1992). Map-makers have long understood the ways in which their craft is one of constructing a persuasive and useful pictorial representation of spatial relations. In this most basic of senses, by selecting some features and ignoring others, maps act like cultural expressions indicative of the society that produced them (Aziz 1978: 50).

These questions have posed fundamental and, I think, intractable problems for cartographic theorists. Certainly not all cartographers accepted such functionalist models of cartography. Many took exception to this way of seeing maps, arguing instead for the importance of semiotic (Schlichtmann 1985, Wood and Fels 1986) cognitive (Petchenik 1983) or cultural (Guelke 1977, 1981) approaches to maps. But even in these reworkings of the meaning of maps, psychologistic models and technical–instrumental understandings of meaning were common. While the limitations of a communication model for understanding map-making and map use soon became apparent, subsequent attempts to model the mapping process quickly approached the baroque. The flow of information was quickly rendered in mathematico-psychological terms as transformations in the flow of information, the techniques of the transformations, and the effectiveness of the map in regard to them (Robinson *et al.* 1984). But, what Bruno Latour has called 'the modern settlement' (the commitment to a binary logic of society–nature and representational logics in politics and science) remained at the heart of cartographic thought. Cartographic representation continued to be conceptualized as the technical transfer of real-world information to users within this modern settlement. It left cartography with the Kantian dilemma of how it knows the world and how it can represent that 'real' world adequately to control the misreading of map users. That is, the management of choice, distortion and error – fundamental to any cartographic representation – became its problem.

As Jan Broek argued (1965: 64), it was the very craftsmanship and persuasive quality of maps that meant that map users have often overlooked the actual practices of map design and map-making. Unlike the author of a written text, the cartographer cannot express the limits of technique in the map itself. The lack of cartographic 'buts' and 'ifs' gave the cartographer 'much less leeway' to remind the map-reader of the interpretative nature of the mapping process, and, as a result, the map-reader easily falls into the habit of seeing 'the map as a precise portrayal of reality' (Broek 1965: 65). This easy tendency to see maps as naïve representations of reality has also meant that the map has been easily adaptable to nationalistic and propagandistic purposes. It has been the very skill of crafty transmutation in which the cartographer translates lines and shadings into worldly reflections that has led to an easy acceptance of naturalism (the 'mirror of nature') and, in the process, provided opportunities for charlatans to propagandize the map. It is to this issue that we now turn.

SECOND CRISIS OF REPRESENTATION: JOHN KIRTLAND WRIGHT AND THE SUBJECTIVE NATURE OF MAPS

It was precisely this emerging scientific notion of maps that prompted John Kirtland Wright to write his classic essay: 'Map-makers are human: comments on the subjective in maps'. First published in wartime conditions in 1942, the essay was concerned with the emergence of propagandistic cartographies of various kinds. But the essay has rarely been read in this context, partly because of the circulation of it through his collected works, published in 1966, and partly because of the ways in which his geometry of modernity (a binary of objectivity and subjectivity) and his moralist tones were so readily adapted to anti-political post-war discourses.

In the essay, Wright provided a spirited defence of the role of the subject in constructing and reading maps against the then emerging empiricist and naturalizing tendencies in geography and cartography. Maps, he began, 'are drawn by men and not turned out automatically by machines, and consequently are influenced by human shortcomings' (Wright 1966: 33). Like Broek before him, Wright argued that it was precisely the 'trim, precise, and clean-cut appearance that a well-drawn map presents' (p. 33) that lends to the map an air of scientific authenticity and a persuasive character that reaches beyond the technical limits of the map itself. The map leaks as a tool so that '[w]e tend to assume too readily that the depiction of the arrangement of things on the earth's surface on a map is equivalent to a photograph ... The object before the camera draws its own image through the operation of optical and chemical processes. The image on a map is drawn by human hands, controlled by operations in a human mind' (Wright 1966: 33). 'Every map is thus a reflection partly of objective realities and partly of subjective elements ... No map ... can be wholly objective' (Wright 1966: 33).

Wright immediately pulls back from this distinction between the objective nature of photography (and the representational economy it suggests) and the subjective nature of cartography. Instead, he turns to what he calls the subjective and objective elements in maps and photographs. 'Even a map of an imaginary country is objective, in the sense that the mountains, roads, towns, and so on that it pictures were suggested by corresponding objective things in the real world' (Wright 1966: 34). For Wright, mapping is a 'Mirror of Nature', but a mirror whose images are occasionally fogged and distorted by the subjective elements of the map-maker and user: 'the maps produced by government surveys or made in the field by explorers are more or less directly copied from nature ... Many maps, however, are not drawn from nature but are compiled from such documentary sources as other maps, surveyors' notes and sketches, photographs, travellers' reports, statistics, and the like. As these sources are themselves man-made, the subjective elements they contain are carried over into the maps based on them' (Wright 1966: 34). In this view, objectivity derives from closeness of observation, in which direct access to the reality of nature is given to the cartographer who can then copy its form. With increasing distance from nature, greater levels of subjective judgement are introduced and these in turn require consideration of the mental and moral qualities of the cartographer and map user. The recognition that map-makers are human requires an attention to questions of 'scientific integrity, judgement, consistency, progressiveness, and their opposites' – a thoroughly modern, American liberal economy of science – a second crisis of representation – has been put in place that will frame the moral economy of geographical discourse for the rest of the century.[9]

THIRD CRISIS OF REPRESENTATION: DISTORTION, ERROR AND PROPAGANDA MAPS

The third crisis of representation with which I want to deal arises from the selective interests that shape all maps. It has to do with the ways in which modern cartography has dealt with distortion in its two linked meanings: error and deception. I focus on the ways in which the objectivist claims of scientific cartography have been lodged against the treatment of propaganda and popular maps. In dealing with this issue, propaganda and popular maps have been marginalized from the cartographic canon, variously referred to as a form of 'graphicacy' akin to literacy and numeracy (Balchin and Coleman 1965/1966), a form of 'cartohypnosis' (Boggs 1947), as 'magical' (Speier 1941), as 'weapons' (Weigert 1941; Herb n.d.), as 'persuasive' (Tyner 1982; Herb 1989), and as a form of 'propaganda tool' (Burnett 1985; Herb 1989; Pickles 1992). In ways that should be unsettlingly familiar to the cartographer, the propaganda cartographer is seen as one who deliberately selects information to support an argument, distort information, and display it in ways that seek to persuade the map-reader of a particular viewpoint. The propagandist structures the production of the map for maximum visual impact as a calculated exercise of 'persuasive cartography' (Ager 1977). That is, propaganda aims at persuading large groups of people to believe something or act in a way that they would not, in the normal course of events.

Propaganda techniques are, then, techniques of persuasion that may fail to abide by established and accepted norms of accuracy and truth. They may seek to manipulate relationships in order to persuade people about a particular claim to truth. But they might just as well deploy truth claims, accurate information and careful argument to make their case. For example, Lord Northcliffe's observation after the First World War that '[t]he bombardment of the German mind was almost as important as the bombardment by the cannon' has since been taken to heart by many propagandists and pundits. Hitler is reported to have argued that 'Propaganda consists in attracting the crowd, and not in educating those who are already educated'. It must be addressed to the emotions, not to the intelligence; it must concentrate on a few simple themes; and it should be presented in black and white. It has little to do with the truth and more to do with historical necessity. But to achieve political and military success the truth was to be strategically deployed, especially in maps. In this strategic thinking, lies were too easily shown to be false and were therefore ineffective. Instead the nucleus of truth or falsity was to be hidden by veils of interpretation, providing a channel of escape if anyone questioned the truth of propaganda (Thomas 1949: 78). Prior to the Second World War, Karl Haushofer had attempted to deploy these ideas in propaganda or suggestive cartography that would transform geopolitics into 'a dynamic Weltanschauung to further the expansive claims for Lebensraum of Germany' (Weigert 1941: 529). Such dynamic suggestive maps relied on the strength of the initial idea and the use of symbolism – the new cartography was to be visually violent – to accost the map-maker and to present a clear message. Often such images are hardly recognizable as maps or the map is only part of a collage of images that wilfully exploit the inherent limitations of maps to distort and exaggerate (Quam 1943: 21). A particularly clear example is a map dropped on the Allies at Dunkirk during the Second World War (Figure 2.2). The map depicted the position of the

Camarades!

Telle est la situation!
En tout cas, la guerre est finie pour vous!
    Vos chefs vont s'enfuir par avion.
A bas les armes!

British Soldiers!

Look at this map: it gives your true situation!
Your troops are entirely surrounded —
        stop fighting!
Put down your arms!

Figure 2.2 Map-poster dropped by German aeroplanes to Allied troops in Belgium while they were fighting, *c.*25 May 1940 (*The Belgian Campaign and the Surrender of the Belgian Army.* The Belgium American Educational Foundation, New York, 1940, with permission)

troops as hopeless. They were shown to be completely surrounded, with little hope of escape and nowhere to escape to: the Allies were surrounded, the Germans were on the move (indicated by the use of bold arrows throughout occupied territory). The technical manipulation of the visual field of the map made the call for men to lay down their arms appear reasonable in such an island of desperation. Hope was removed visually from the map by the failure to show the south-east coast of England 30 or 50 miles across the Channel.

Such notions of propaganda are, of course, already centred on an unexamined boundary between 'truth' and 'falsity', an unstable boundary at best; one that Gramsci (1981: 80, n. 49) sought to reconfigure in his analysis of hegemony:

> The 'normal' exercise of hegemony on the now classical terrain of the parliamentary regime is characterised by the combination of force and consent. Indeed, the attempt is always made to ensure that force will appear to be based on the consent of the majority, expressed by the so-called organs of public opinion – newspapers and associations – which, therefore, in certain situations, are artificially multiplied.

Institutions concerned with the process of establishing hegemony all too readily and easily use such techniques to capture the discursive field and reconstitute the discourse of the age and the place, and it is this discursive capture that cartographers such as Monmonier (1989, 1991, 2001) and MacEachren (1994, 1995) have attempted to problematize.

The map has been an archetype for such kinds of hegemonic projects in the historical construction of the nation-states, where it has been an essential tool in territorializing the state by extending systems of policing and administration, and in establishing a sense of national identity at home and abroad (sometimes in the face of explicit internal disunity or rebellion). The state must consistently attempt to capture the discursive and ideological field not only through the more obvious organs of public opinion, but also by the appropriation of space (and the map) to its purposes and by the symbolic constitution of mapped space as national space (Figure 2.3). Here the link between map and symbol becomes clearer. The territorialized state, symbolized in unity under the sign of the *Leo Hollandicus*, is rendered as integral, unified and powerful. The unity of President–government–territory–nation is captured in the unusual 1912 Roosevelt map of the United States (Figure 2.4). Here complex (unrepresented) histories of texts are evoked as the anthropomorphized map unifies people and land; the resultant unitary state – the nation-state – is personified through a single figure, a personification that itself evokes multiple other texts. In this sense, meaning is produced through the invocation of chains of national–patriotic–territorial signifiers. The map evokes, not represents, these unrepresented signifiers.

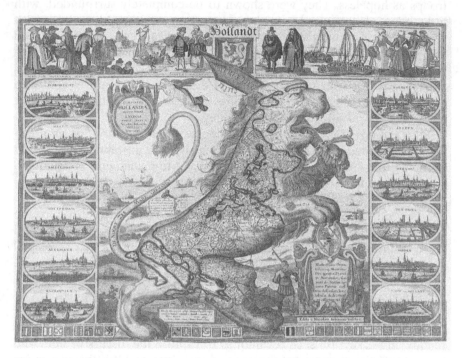

Figure 2.3 Leo Hollandicus, Claes Jansz, Visscher, 1648. The territorializing of the
geo-body of the nation-state depended, in part, on the map. In the *Leo
Hollandicus* map, the Seven United Provinces of the Low Countries
were depicted as a lion, symbolizing an imagined and hoped for
community of unity and power

It is perhaps the association between propaganda and the politics of
totalization in the twentieth century that has diverted attention from the
ubiquitous uses of propaganda generally and propaganda maps specifically
in all sorts of projects that seek to territorialize identity and foster hegemony: national mapping programmes, commercial advertising, the everyday work of public institutions and the construction of our own 'enemies'
(see Zizek (2001) for a more extensive treatment of this politics of purification and exclusion). In my view, this association and demonization of
propaganda maps has led to a narrowing of the cartographic canon and the
bolstering of a technicist and instrument understanding of representation.
Propaganda maps and popular maps have been treated as exceptionalist
and they have been exempted from theories of maps and cartography.

In their rush to create a science of mapping, post-war cartographers
have too quickly forgotten the lessons of their war-time colleagues.
Indeed, perhaps the clearest statements of the map–propaganda map relation was provided by the very cartographers who were engaged in combat-
ing German attempts to use propaganda maps during the Second World

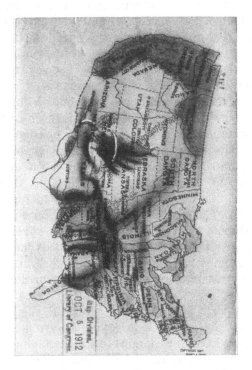

Figure 2.4 Roosevelt Map of the US (Library of Congress, Washington, DC)

War. For Weigert (1941: 5301) the map was a double-edged weapon: 'in unskilled hands it easily becomes a subject of ruthless and stupid propaganda. But in the hands of the expert who knows the rules of the war of words as well as those of modern cartography, it is a good weapon ... it can bring hope to the suppressed nations and fright to their suppressors. And here too, the attack is the best defense.' Here, science, accuracy and truth were to be mobilized in defence of democracy. Map-makers 'must strive to make their maps accurate and in harmony with the democratic ideals of our cause' (Quam 1943: 32), a task made more urgent by the fact that as: 'global war progresses the harder it is for even the generally well informed and earnestly interested citizen to keep track of all its rapidly changing aspects and the more difficult grows the task incumbent upon the various media of information' (Soffner 1942: 465).

Such lessons of persuasive cartography were not lost on commercial designers, who have always seen the benefit of the recomposing images, arraying them in series, using incomplete forms, and encouraging viewers to 'fill in' the boundaries with broader cultural imaginaries, techniques that are currently being developed with immense sophistication in American television commercial and public service advertising (Levi jeans, anti-smoking ads, and 'just say no' ads) (McDermott 1969). In each visual 'clip' (the commercial equivalent of the soundbite) sustained plot lines appear to be absent. Instead, the narrative structure is built temporally with clips building on each other to form a coherent and often powerful composite impression. Cold war geopolitics have long been fostered through similar partial, and in themselves often meaningless, visual 'clips' drawing on historical and often biological analogies to produce a kind of temporal montage. Some of the most sustained of these images in cartography are those that evoke imperial metaphors of reach, expansion and power. The Russian bear was a form repeated in various guises, playing off earlier uses of the image and depending for its impact on them. Extremely influential elaborations of this image were to be found throughout the cold war from atlases using Mercator projections to render superpower status, to R.M. Chapin's communist contagion map in *Time* magazine for 1 April 1946 and Red China in 1955, to the covers of Defense Department codebooks. In the map 'Red China', Chapin's careful selection of shading and symbolism permitted him to illustrate the red menace reaching round the Chinese mainland (see reproduction in Pickles 1992b). The threat from the Soviet Union, North Korea and Vietnam is visually focused first on China, but then on to the island of Okinawa, where stands the stars and stripes. Moreover, the gross exaggeration of the Himalayas closing in on the margins of China emphasizes its isolation from the West, whose surrogates India and Pakistan are shown in the recessive colours of light green. The message is strong but not obvious. The whole map is a study in suggestion, in which cartographic techniques are used to depict a particular situation in such a way that both the intrinsic meaning and the suggested meaning resonate with other texts and images beyond this single map.

On occasion, the globe and the map have become such successful symbolic images that their 'shadows' can be presumed in images that contain no map form at all. In geopolitics, one especially rich and evocative example of this adaptation of the globe has been the motif of the spatial reach of Empire, symbolized by the arms of the spider or octopus (Figure 2.5). Here the historical repetition and reworking of the same image has permitted the cartographic specification to be removed. No map or country location is given, but one is presumed. At this point the map and the cartoon fuse. The map form is present as a kind of technical and historical memory; an unacknowledged absence that is constitutive of the image and essential for any interpretation. In practice *all maps* exist within similar unacknowledged contexts of other maps, symbols and meanings, and any theory of maps must find a way to deal with the work done by such absent contexts.

In each of these examples meaning arises from the merging of multiple horizons, some directly represented, some evoked, some presumed. As Weigert (1941: 528) suggests:

Figure 2.5 Cartographic tropes of imperial power and reach

it is surprising to see that we are not all conscious of the important part which the map and the art of map-making plays in the process of creating a new conception of the world. We simply rely on maps as if they were facts in this transformation of thinking and seeing. The astounding observation that, in the discussion of the vital problems of the day, the maps as they are presented to us are being taken as stable and indisputable facts, as mere tools which do not themselves reflect aims and opinions of their creators – this naive confidence in the truthfulness of the map indicates that many of us are not aware that maps are weapons. Like the written and spoken word, like photographs and cartoons, the map has become a psychological weapon in a warring world where the souls of men are as strongly attacked as their lives.

And this is surely the point: mapping is an interpretative act, not a purely technical one, in which the product – the map – conveys not merely the facts but also and always the author's intention, and all the acknowledged and unacknowledged conditions and values any author (and his/her profession, time and culture) bring to a work. Thus, like all works, the map carries along with it so much more than the author intended. Also, like any text, the map takes on a life (and a context) of its own beyond the author's control. The map is a text, like any other in this regard, whose meaning and impact may go far beyond the limits of technique, the author's intention, and the mere transmittal of information. Thus, the perception of graphical images is not a purely psychological reception of information but a complex social play of images present and absent, in the context of other symbolic, ideological and material concerns. All cartography operates within and makes use of such unacknowledged preconditions and more or less accepted symbolic forms and mapping conventions. The impact of these techniques and effects are only clearer and sharper in propagandistic texts.

Certainly, modern cartography has done little to elucidate what might be the social, historical and technical metadata needed for an understanding of the work being done by a particular map. Strangely, while the question has been an important one in geographic information systems and new digital mapping, and while specification of projection, scale, legend and the date of the map's production has always been important to cartographers, this kind of contextual metadata about the production, circulation and consumption of maps in the plural and in their historical specificity has rarely been attempted. Instead, the individual map has been the locus of attention and the frame of analysis; unacknowledged context and histories have been dealt with in terms of the history of maps, not as a fundamental technical and ethical issue of map production and use.

It seems to me, at least, that it is precisely this failure to deal with the hidden presences and the metaphorical and symbolic complexities of maps that has produced such a limited reading of propaganda maps (and maps generally). While cartographers are seen to present information accurately, comprehensively, with a balanced design, and without favouring one side of an issue, the propaganda cartographer is seen 'to produce a map which has visual impact and is not only believable, but goes a stage further – is convincing' (Ager 1977: 1). Propaganda maps are problematic because the cartographer has used the wrong method and has 'failed to communicate correctly with the user' (Ager 1977: 14). The cartographer's colour choice, use of lines, orientation of north to the top of the page, and choice of material which will appear at the centre of the map, as opposed to at the edges, are all elements that are 'extraneous to the scientific purpose of the map' (Speier 1941: 313). The propagandist exploits these elements: 'The propagandist's primary concern is never the truth of an idea but its successful communication to a public. Geography as a science and cartography as a technique become subservient to the demands of effective symbol manipulation' (Speier 1941: 313).

Judith Tyner (1982: 2) has suggested the name 'persuasive cartography' to distinguish such propaganda, suggestive, advertising, journalistic and subjective cartography from other forms. Persuasive cartography is a 'type of cartography whose main object or effect is to influence the reader's opinion, in contrast to most cartography which strives to be objective'. Persuasive cartography thus seeks to manipulate symbols in order to influence some group about the value of some idea, opinion or action. But as Ager (1977) pointed out: 'in reality there is not a clear division between "Propa-

ganda" cartographers and "Perfect" cartographers, but both are at opposite
ends of a spectrum in which all cartographers fall, and their positions vary
in accordance with the production of each map.' But while such distinctions
may be difficult to draw in practice, it is the very distinction between objective cartography on the one hand and biased or propaganda maps on the
other that may be the problem. What, for example, can we say about maps
such as those that represent the 'earth from the South' or 'the Australian's
view of the world'? In such maps, the techniques of modern cartography
(and often very finely crafted maps) are used to dislodge a particular hegemonic orientation that has been standardized historically. Is this to be characterized as a propaganda or a political map, but the more standard
northern orientation of other modern maps is not simply because the latter
has been accepted as the norm? What are the limits of standards and
norms, and when is a norm itself a form of propaganda? Is the distinction
between propaganda and scientific cartography dependent on specific
moral and historical judgements about accepted practice? Does scientific
cartography not use the arts of persuasion, distortion and aesthetics? What
can we say, in this context, about the cartographies of reconstruction drawn
by Hans Speier for national socialist Berlin (Figure 2.6)? Surely such
technically accomplished maps for the new city of Berlin must be contextu-
alized within Speier's own claims for cartography and the relationship
between cartography and the planning of a post-war national socialist city.
That is, the scientificity of Speier's maps requires a historico-political analysis of his cartography. Is it really sustainable to claim that the style, form
and underwriting of the maps produced by classical and modern cartography have not been shot through with equivalent (albeit less abhorrent and
violent) political and social interests of one kind or another?

By not paying sufficient attention to their own crafty skills of transmutation and by tirelessly seeking to turn away from the interpretative nature of

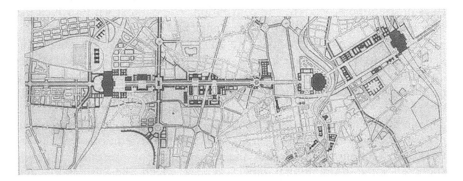

Figure 2.6 'Geography as a science and cartography as a technique become subservient to the demands of effective symbol manipulation' (Hans Speier
1941: 313). Albert Speer's map for the reconstruction of Berlin (Library
of Congress, Washington, DC)

their Merlinesque constructions, scientific cartographers have found it very difficult to explain the difference between their own magic and the conjuring tricks of those who would use this magic for militaristic, propagandistic or commercial ends. By making scientific cartography into a technical enterprise and by rejecting (or overlooking) its magical and hermeneutical practices, cartographers have grappled with error and distortion in only technical terms. In this scientistic view, the management of error has been rendered only in terms of technical error, malicious intent or the limitations of the untrained. They have, as a result, also made maps much less interesting!

Here the three crises of representation (objectivism, liberalism and subjectivism) and a commitment to transparency coalesce to frame a contemporary theory of scientific cartography, one that locates theory and practice firmly within the domain of technical expertise. Cartographic technique is seen as an ongoing approximation to the real, presupposing a correspondence or representational theory of truth. The distinction between fact and fiction is mirrored by the separation of the good cartographer and the propaganda cartographer (the latter being banished from the halls of science). The ideological is expelled, but from a world that disavows its own ideology, its own history, and its own commitments to transparency. Science is seen not as a persuasive enterprise but as a claim to true knowledge. A good map is one in which the image received by the map user corresponds to that intended (inscribed) by the map-maker and where the image inscribed (and received) is an accurate representation of the real world. Map-making and map-reading are seen to involve the straightforward transmission of information in a philosophically and practically unproblematic manner. In particular, while cartography always does seek to persuade, to convince or to argue, it does so without selecting its techniques for purely visual impact; in the choice of subject matter, what is centred on the page, what is consigned to the edge of the map, and which scale and projection shall be used, the cartographer is guided by rules of scientific procedure and convention. The context within which the map is interpreted is restricted on pragmatic and technical grounds.

For some, the patently inadequate boundary between science and non-science and between technical error and intentional error inscribed in this understanding of cartography suggests an alternative theory of maps. One resolution might be to assert that all maps are propaganda maps. But this too fails to deal with the problem. Since all maps are constructed images, and since all images are interpretations of a particular context, we gain little by merely repeating that maps are both interpretations and distortions. We remain caught within the metaphysics of presence that presupposes some foundational object against which the distortions and interpretations can be measured: that some interpretation-free image could be produced that does not distort the world. At this point, liberal cartographic theory merely asserts the difference between its legitimate and illegitimate children. Recognizing that all maps distort, cartographic theory insists that what matters is the intention behind the construction of the map and the use to which a map is put. For Monmonier (2002: 640) 'it's the situation that makes a technology good or bad'. But, what is meant by 'the situation' and what are the historical, geographical, and social chains of causality and effect bound by 'the situation'? It is clear that for Monmonier 'the situation' means specific spatio-temporal practices, immediate uses, possibly bound together by repetitions of inappropriate practice. In this view, maps are neutral until activated within a specific context.

Karl Figlio (1996: 73–6) has attempted to wriggle free from this liberal impasse and the metaphysics of presence it presupposes, arguing that mapping is a representational act that both presents the world and annuls it at the same time. 'Every mapping into geometrical spaces – every picturing leaves a gap between what was present in emotional space ... And what appears in the mapped space' (p. 75). Mapping is therefore the building of repression. Like Mary Shelley's Frankenstein (perhaps the extreme example of scientific mapping), the creation is a monster, shunned and forced to live ignored until finally it takes its revenge. For Figlio, science acts in this way: it concocts nature, reduces its dimensionality, represents it in non-contradictory, bounded form. But this project of visualization is predicated on both repression and leakage; the monster emerges as an actor in its own right, no longer as representation but as the return of the repressed. In this sense propaganda maps function as the repressed creation of scientific, objective cartography; a monster created and unleashed by its own logics and practices. If, as Bruno Latour (1999) suggests, we simply abandon our commitment to both objectivism and naturalism, as objectivist epistemologies breathe their last breath, such monsters may just shrivel up.

BEYOND OBJECTIVISM AND RELATIVISM

In 1968 Juergen Habermas published *Knowledge and Human Interests*. In it he showed so well how the hubris of modern objectivist epistemologies could not be sustained. Knowledge claims were always embedded in forms of social interest; not only interests of specific social groups but also broader epistemological interests in technical, interpretative and critical knowledge. Such claims found resonance in geography. For Zelinsky (1973): 'a map ... has meaning only as it relates to other aspects of an interlocking communicative structure' and 'can only be understood as one of several elements in a complex series of transactions, in constant state of flux, involving: (i) an objective reality "of some sort"; (ii) explorers or observers; (iii) the map-maker; (iv) the document; and (v) the map-reader or, more realistically, a community of map-readers.'

With Brian Harley's later work, the study of maps and cartography as products of human endeavours, social interests, and institutional powers became an established and legitimate area of inquiry.[10] Both in terms of the specific claims made in Harley's writings, but also in terms of the ways in which 'Harley' and 'deconstructing the map' have entered the lexicon of critical human geography and cartographic studies, the study of maps as other than simple iconic devices or complexes technical products has gained widespread acceptance. As Harley suggested in his introduction to Volume 1 (Harley and Woodward 1992: 1), in making the principal concern of the history of cartography the study of the map in human terms,

> the *History of Cartography* is concerned, as far as possible, with the historical process by which graphic language of maps has been created and used. At once a technical, a cultural, and a social history of mapping, it rejects the view of a historian of discovery who wrote that 'cartographical studies do not come within the sphere of social history' ... On the contrary, it favors an approach that is potentially capable of exploring the behavioral and ideological implications of its subject matter.

For Harley (and for geographers and cartographers since), the map was a social product and a social actor, a product of and embedded in complex networks of social relations and interests. Like any other technology and product, the map must be interrogated in its social contexts of emergence, dissemination, and use. But for Harley (1990: 1) the writing of a social history of cartography as a set of practices was even more crucial. The crisis of representation is also a crisis of democratic practice and ethics in which technical knowledge (in particular digital geographical information systems) displaces more accessible hard-copy maps that have, for generations, allowed a certain kind of public practice and exercise of civil society in the face of power. The result is the need for a strong debate about the ethics of representational practices and cartographic goals. In this debate, Harley (1990: 2) sought to foster

> a *public* agenda that seeks through an open debate to extend cartographic consciousness beyond a narrow concern with 'accuracy' or 'utility' as the sole ethical yardsticks. It will become clear that I believe that our discourse about maps, whether historical or modern, should be more responsive to social issues such as those relating to the environment, poverty, or to the ways in which the rights and cultures of minorities are represented on maps.

Harley began his 'Deconstructing the map' with a basic question and surprising answer. He asked, what is a map? And he answered, 'cartography is seldom what cartographers say it is' (Harley 1989b: 1). For most cartographers,

> [t]he object of mapping is to produce a 'correct' relational model of the terrain. Its assumptions are that the objects in the world to be

mapped are real and objective, and that they enjoy an existence independent of the cartographer; that their reality can be expressed in mathematical terms; that systematic observation and measurement offer the only route to cartographic truth; and that this truth can be independently verified.

But, for Harley, maps were always social creations, embedded in networks of social relations and interests, reflecting them intentionally and unintentionally. The hidden agenda of mapping (including cartography's modern claim to accuracy of representation) is precisely what makes them interesting and problematic texts, first in terms of the silences of maps (those elements of the landscape that are omitted) and second (and often related to the first), in terms of the implicit and explicit authoritarian nature of the map as a tool of power (of the state, military or capital) (Harley 1989b: 14). In particular, this authoritarian nature of the map had to do with the service to the state provided by cartography as a power–knowledge: 'As cartography became more "objective" through the state's patronage, so it was also imprisoned by a different subjectivity, that inherent in its replication of the state's dominant ideology' (Harley 1988a: 71). As a result, the history of cartography is to be both an uncovering of hidden agendas, silences, elisions *and* ideology critique. It is part genealogy of the power–knowledge that cartography constructed around its practices, and part a deconstruction of the demarcations and delimitations that found cartography's own claims to objectivity and science (see Editorial comment 1992: xx).

MAP-READING

As we have seen, traditional theories of maps and map-reading are of little use to us when we push them to incorporate propaganda maps and the broader socio-cultural contexts within which maps crystallize determinate meanings. Without the foundation of an unproblematic theory of representation to fall back upon, cartographers retreat to the position that all maps are distorting and hence all maps function as propaganda maps. But as we have also seen this is merely to sidestep the issue and raises other serious questions about the sorts of claims we can make and the work we can do. Both approaches do not adequately address the textual qualities and commodified nature of maps. In this section, I unpack the textuality of maps in three ways: the world and the text, the text in a text, and the analysis of the work itself.

We encounter ambiguity the moment we ask: what is the content of a graphic image? Clearly it is the real world, the real situation, the landscape, the scene. The map-maker reduces this object-field according to established principles of objectification, abstraction, reduction and idealization to create the map. In this sense all maps are thematic abstractions involving reduction of one form or another (see Harbison 1977 quote at the beginning of this chapter). In ways quite different from the photographic image, however, this reduction is a particular form of transformation. In order to move from the real situation to the map it is necessary to divide up this reality into units and to constitute these units as signs substantially different from the object they communicate (as the Harbison quotation suggests). The map is thus a coded message whose relationship to the object-world it evokes is particularly complex. While photographs are also complex recodings of visual schemas, objects and contexts, the map requires additional projections, symbols and codings.

What is the nature of these codings? The map is a message. As the previous discussion suggests, cartographers and geographers have traditionally taken this message to involve a source, a medium and a receiver. The source is the cartographer (and his or her body of received techniques and style), the medium is the map (and the often ignored immediate contexts within which the map is embedded) and the receiver is the map-reader (as a public or professional 'readership'). But as we have also begun to see, this view construes the map too narrowly. It ignores the other texts within which the map is itself embedded and with which it is codetermined. It ignores the context into which the map is projected and of which it is a projection. It one-sidedly places emphasis on the intended message and fails to consider possible unintended meanings. Finally, it has no way of accounting for the ability of graphic images to conjure up other texts (maps, photographs, books, etc.) and embed them in any reading of their own codes. By way of illustration, let us ask, what is the medium of the map's message? In the communication models discussed above, the medium is the map. But how can this be? The medium is the report, the article, the book, the magazine, within which the map appears. More precisely, as Barthes (1978: 15) says of the photograph, the medium is 'a complex of concurrent messages with the photograph as centre and surrounds constituted by the text, the title, the caption, the lay-out and, in a more abstract but no less "informative" way, by the very name of the publication.'

We are faced with layers of textuality: the map itself, the immediate context of the map (its caption, the chapter and the work of which it is a part) and the wider context of the map (the opus of the individual cartographer or school, the opus to which the text itself belongs, the socio-cultural context of the work). But although the map is an embedded figure, the map is also an object that has a structural autonomy independent of both its production and its use, and thus requires an analysis of the work itself. This will not be definitive, but will always have to be situated alongside sociological, historical and geographical analyses of text and context; of production and use. Even an analysis of the work itself cannot divorce the map entirely from its context, for the map is not an isolated object. It has a title and fits within the body of a text, along with a set of other maps, or, if it is a single map, it is framed and displayed in some manner.

In terms of the internal construction of the map, the message of the map is carried by at least three different structures, one of which is graphical, one is mathematical, and one is linguistic. Yet the consideration given to the linguistic components of the map has been mainly restricted to the design and effectiveness of the graphicality of lettering (size, print style, placement). While in other graphic forms (photography, painting) the graphical and linguistic elements are complementary, in the map they operate almost uniquely as inseparable from each other. This inseparability is also typical of certain forms of advertising, poster art and modernist art forms such as Dadaism. Here the linguistic elements are embedded within the image, not incidentally but as intrinsic components of the whole picture (Figure 2.7).

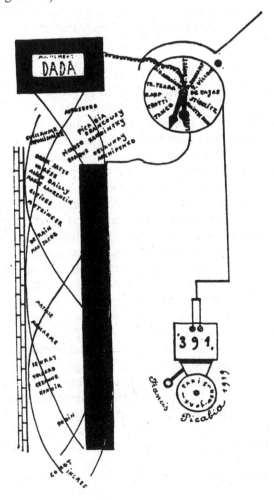

Figure 2.7 Dada Movement, 1919, by Francis Picabia (with permission of Museum of Modern Art, New York)

In describing the linguisticality of the title of the photograph Barthes (1978: 16) says:

> The two structures are co-operative but, since their units are heterogeneous, necessarily remain separate from one another: here [in the text] the substance of the message is made up of words; there [in the photo-graph] of lines, surface, shades. Moreover, the two structures of the message each occupy their own defined spaces, these being contiguous but not 'homogenized', as they are for example in the rebus which fuses words and images in a single line of reading. Hence, although a press photograph is never without a written commentary, the analysis must first of all bear on each separate structure; it is only when the study of each structure has been exhausted that it will be possible to understand the manner in which they complement one another.

But this position is untenable for the theory of maps. In the map, the symbolic graphic image is embedded in a written text (a paper, a book, an atlas) and rarely has an existence beyond the body of the text and the discursive aims of the research of which it is a part. Moreover the symbols and words in the map are interbedded: the names of places, features and other descriptors are integral to the visual image, and call for a special form of construction and present specific difficulties for analysis. In particular, such interbedded texts (maps, poster, commercial and Dadaist art) are correspondingly much closer to the tract: the commercial, political poster or artwork. It is in this intersection that much of what has been called propaganda mapping arises. The issue becomes clearer on closer analysis.

For Barthes, all 'imitative' arts comprise two messages: a denotative message and a connotative message. The realist painter and photographer stake their reputations on their work being predominantly denotative, in the sense that the representation of the objects is a representation in which the objects represented are objects from the world, without obvious transformation. Where such transformations are integral to the image they build on some expectation of verisimilitude. Traditional cartographic theory presents the map as a purely denotative message. But, as we have already seen, in the mapping process objects are transformed and reconstituted as signs and symbols substantially different from the objects they communicate. That is to say, the map is a coded message.

There is another important sense in which the map differs from the photograph. In most photographs (except where the object of the photograph contains language), the caption constitutes a parasitic message that adds to and circumscribes the meaning of the photograph. In the map the issue is more complicated. The caption here is also a parasitic, albeit essential, part of the map. First, it merely illustrates the image, often through a repetition of the more obvious content of the map image itself. Second, 'the text loads the image, burdening it with a culture, a moral, an imagination' (Barthes 1978: 26). The caption also reinterprets the map and points us to specific or specified meanings; the caption circumscribes our reading of the map. Third, the map image itself is also linguistic. Here the interplay of codes and words constitutes a distinctive image form in which the message is achieved largely in terms of the interplay and duality of graphic and linguistic meaning.

The transmission and reception of the map image are not the straightforward, linear process presumed in the communication model. The coded image (the map, linguistic and graphic) is also connotative. Through the fusion of horizons between the reader's world and the world of the map (and the map-maker) the map connotes a variety of meanings. Thus the reading of the map is always historical and 'depends on the reader's "knowledge" just as though it were a matter of a real language, intelligible only if one has learned the signs' (Barthes 1978: 28). The map is a purposive cultural object with reasons behind its construction and values associated with its reading. To suggest otherwise is to fail to see its status as made object. The map is always and necessarily an expression of an idea. In mediating the transformative processes of abstraction, reduction, thematization and idealization, the cartographer selects, sifts and emphasizes this or that aspect of the world under consideration, and articulates an image in the rebus linking graphic and linguistic codes.

If Barthes's distinction between denotative and connotative meaning allows us to open a first step in the analysis of the work, further reflection on the distinction forces us to abandon it. The experience of modern art forces us to rethink the very nature of this distinction. John Berger (1965: 55) explains this change in showing how the revolutionary vision of Cubism arose out of an inheritance passed to us from the nineteenth century:

> Nature in the picture is no longer something laid out in front of the spectator for him to examine. It now includes him and the evidence of his senses and his constantly changing relationships to what he is seeing. Before Cézanne, every painting was to some extent like a view seen through a window. Courbet had tried to open the window and climb out. Cézanne broke the glass. The room became part of the landscape, the viewer part of the view.

Thus the challenge of modern art and modern science is to work through the implications of accepting the inevitability of our participation. For Heisenberg (1959) this meant that 'Natural science does not simply describe and explain nature; it is part of the interplay between nature and ourselves; it describes nature as exposed to our method of questioning.' Failure to come to terms with this participation has serious consequences. It was the power of Cubist painters before 1914 that they were able to link Courbet's materialism with Cézanne's dialectical view of the image. But one without the other would have led and did lead to a sterile art. Materialism became literal and mechanical. An ungrounded dialectical view became disembodied and overly abstract. The danger for a theory of maps/texts is obvious. A representational view of the image divorced from an investigation of the role of the one who constructs the image becomes literal and mechanistic. Conversely, overemphasis on the viewer and the viewer's responses becomes idealistic and equally reductionistic.

The question arises, then, where will we find a theory that can deal with maps as texts, without reducing all map forms to forms of propaganda maps? A preliminary answer might be found in a reconsideration of critical hermeneutics.

HERMENEUTICS

In the nineteenth century, the clock became the metaphor for mechanical approaches to the social sciences. In the twentieth century, the electrical circuit became the metaphor for systems approaches in the social sciences. These and related metaphors have left lasting impressions on twentieth-century social science, Yet in the second half of the twentieth century a new metaphor – the text-metaphor – emerged as the template for understanding and framing social life. In this period the text-metaphor has colonized certain domains of study – painting, film, landscape and most recently social life.

Extensive use has been made of the analogy of reading and the text-metaphor throughout the history of modern geography: Sauer's reading of the origins and development of past landscapes from the tracings and antecedents in the contemporary landscape; Lewis's axioms for reading the landscape; Samuels's biography of landscapes; Meinig's symbolic landscapes; Jackson's close interpretation of vernacular artefacts as symbols that reflect broader social changes; Sitwell's equation of elements of landscape with figures of speech; Duncan's studies of the language and semantics of cultural and symbolic inscription; and, of course, the map as an encoded artefact.

How, then, do we read maps, especially those in which problems of interpretation are compounded by distortion, error and lies? More generally, how do geographers read texts? Like the map, the landscape is a particularly good example of a 'text' which has been presumed to require a straightforward literal reading, but which actually poses great problems of interpretation and understanding. Map and landscape each present specific problems of authorship, syntax and structure by which to read (and knowing what not to read), and distinguishing and relating the various levels of determination that historically constituted any particular map or landscape. In the case of the propaganda map an additional problem is always apparent (although it may not always be absent from the land-scape, the film or the novel). From its conception, the propaganda map aims to be a convincing distortion. Hermeneutics is the theory of interpre-tation that deals with problematic texts (their origin, correct ascription, intended meaning, received meaning, etc.). The propaganda map is thus the archetypical problematical text requiring hermeneutic interpretation and provides a potentially good starting point for elaborating the methods of interpretation; philology, hermeneutics and criticism. The previous sec-tions of this chapter have attempted to begin the process of hermeneutic analysis. This section will abstract the lessons and principles of that analy-sis and generalize hermeneutics to all interpretative acts.

Philology places strong demands on the act of interpreting texts (be they poems, landscapes, maps or social actions). Is the text the one it is claimed to be? Is the ascribed authorship correct? Did the text fulfil the role it is claimed to have filled? Is it a coherent whole? What does the text say about its own world? What does the text now mean? What is the relationship between the meaning of a text and the intention of the author in creating it? Given that some of these texts may have been authored by people who are no longer known or who were anony-mous at the time of production, that they may have originated in worlds about which we now know little or nothing, and that only fragments may now be extant, are we really able to retrieve the *mens auctoris* (the author's intention)? And if we are, then in what sense can we claim to have access to the *mens auctoris*? Does the work constitute something independent of and different from that intention? And, if we cut our inter-pretation loose from the author's intention, how do we understand the meaning of a map?

Strict concern for the *mens auctoris* would, of course, place us in an untenable position as social scientists. The antiquarian may claim to bracket his/her present world and become immersed 'fully' in the world of the other, of the past, of the author. This option is not open to the social scientist (nor, practically, even to the antiquarian). We ask questions from the standpoint of the present, and we carry out a retrieval of the author, his/her intentions, and the work, in order to make them meaningful in our present worlds (be they conceptually, temporally or geographically removed). In this interpretative process there is no essential core of meaning or intention in the text to be uncovered. While all texts have an autonomy of their own even beyond the intentions of their author(s), and while the author undoubtedly retains a claim on the surface details of the work: the site, the literal and symbolic content intended, the date of production, the materials and techniques used in production, interpre-tation is always a project of innovation and creation. Neither the content of the text nor the author's intention are fully determinate of the meaning of the text. Instead interpretations resituate the work and rearticulate it in different contexts. The philological concern thus corresponds to a lower hermeneutic, which is concerned to establish a critical edition of a text, to verify that the text is the text it is claimed to be, that it has not been falsified, that it is (or is not) a coherent whole and is not (or is) a pastiche of several authors, that it is authentic and that it is complete. Higher hermeneutics takes as its task understanding the meaning of a text, how it related to its own world (and subsequent worlds in which it has had an existence) and how it is to be related to our present world.

These claims become clearer when we think of symbols in each of two ways: univocal or equivocal. Like symbols in symbolic logic or mathematics univocal signs have one designated meaning. In a limited sense cartographic symbols (church, castle, urban area) have often been seen to be univocal symbols in this manner. However, their correspondence is not of the same kind as that of those symbols of logic or mathematics where the equivalence is complete. In the case of the symbol 'church' the equivalence, as with mathematical notation, is purely formal. Except in this trivial formal equivalence the symbol is actually equivocal: the age, style, denomination and size of the church remain open to interpretation from the context of the whole map and the text within which it is embedded, the choice of warranting the church instead of some other type of building, the privileging of buildings over non-built forms, all require broader social analysis. Equivocal symbols may have several layers of meaning and are a key focus of hermeneutics (Ricoeur 1971).

At least in principle, several conditions guide all interpretations. Interpretation assumes that the integrity of the meaning of the text must be preserved in such a way that meaning is derived from, not projected into, the text. That is, for meaningful discussion about a text an interpreter must first bring him/herself into attunement with the text. This is not a slavish adherence to the text or the tradition to which it belongs. Indeed, as Foucault and others have suggested, the attunement might well be one of seeking the absences, fault-lines, and erasures in the text. But in whatever way the attunement is achieved, the text must be present as a necessary condition for any interpretation.

Of course, the difficulty is precisely where to begin and where to end in reconciling these various responsibilities: how do we understand a text as a whole and how do we understand its parts since all texts have a certain anticipation of their parts from the whole, yet the whole is composed only of parts? Image–event, line–meaning, object–relation, sign–signified, map–world; all are parts–whole complexes of relation and meaning. Part–whole relationships permeate all readings of texts at all levels of analysis and critique, and specifically include:

a   the relationship of the text to its own intrinsic parts;
b   the relationship of language–text–language;
c   the relationship of cultural context–text–cultural context;
d   the relationship of author and his/her world–text as part of this world.

Moreover, since clear and unambiguous texts are indeed rare, all texts must be complemented with suitable assumptions in order for the interpreter to make things explicit that the author and traditions of subsequent reading left implicit. In this way we can say that the interpreter always understands the text in ways different to the author; and there goes the scientific cartographer's attempt to control the process of communicating of meaning through the map! The theory of interpretation as 'the communication of information' predicated on the notion that symbols have determinate, even univocal meanings, that they are 'transmitted' like boxes of chocolates, and that they mirror the world in determinate ways has produced a theory of cartography too one-sidedly concerned with the *mens auctoris*, map objectivity and the technical alignment of the map user with the map-maker. It has centred itself in a technical rationality of cognitive engineering and divorced itself from the broader debates about meaning that might reshape a theory of maps.

THEORY OF MAPS? THEORIES OF READING AND WRITING?

How, then, do we read maps? In particular, how do we answer our initial question: how do we read propaganda maps? Propaganda maps are not a separate category of text and they cannot be accounted for adequately by traditional theories of maps. Instead, an effective critique of the distorting and ideological nature of propaganda maps must be based on a wider conception of what constitutes both propaganda and science. That is, the ideological and propagandistic elements of contemporary 'scientific' maps must also be assessed at those points where the cartographer shares the ideology of his/her age, where accepted practices are founded on particular ideologies, and where unchallenged interests influence the form and content of the theory and practice of mapping. Examples of the ways in which cartography shares and reproduces the values of the age are numerous and some are well known: the continued public use of the Mercator and adapted Mercator projections, the ideological fixation on 'north at the top' maps, and the polite laughter which greets, for example, 'the Australian's view of the world'. Other examples are less well recognized: the focus in western cartography on private property boundaries and lines and the failure to give equal form to public rights of access and usufruct; or the focus of mapping convention on natural and built physical objects, rather than developing universal conventions dealing with symbol, affect or movement.

Interpreting the meaning of maps also requires that other issues be considered. Two symbolic systems are involved: graphical images and writing systems. Not only does the image exist in a reductive relationship to the world, but graphical systems always also exist as interplays between images, linguistic texts and broader social contexts. As a writing system, maps contain within them the spoken and written in a relationship which is never exactly correspondent (i.e., maps 'play' in at least two registers). For cartographers, this complexity of meaning has generally been seen as a technical problem to be either dissolved by careful adherence to established mapping practices or explained in terms of the creative ability of the map-maker. For Thomas (1949: 76):

> Regardless of the objectivity with which they were prepared, a great percentage of existing historical maps present some information which some individuals honestly consider 'propaganda'. For certain areas, the historical issues are so complicated and the record goes back so far that an unbiased map presentation becomes almost impossible.

The issue has only become more complicated not less. Modern cartography is now completely bound to new technologies and practices of computer-assisted information storage and retrieval, graphic display, image production and electro-digital communication. That this has changed both the character of the map and the nature of the map-makers' craft seems undeniable. We must understand what it means for a theory of maps (see Derrida 1981: 13).

In traditional theory, the inscription itself has no intrinsic value, only serving to record a discourse that has already taken place or an idea already formed (either in speech, in the mind of the author or in action). It is therefore testable in terms of accuracy or truth – as an accurate portrayal or resemblance of what is 'engraved on the psychic surface' (Derrida 1981: 184–8). Here inscription is a representation or copy of the *mens auctoris*. This instrumentalist and technicist view of writing valorizes essence over the written form, and is to be overcome by focusing on the exteriority of the written work. In this view, writing is merely the external expression of speech, and writing and speech are merely the external expression of thought (Ulmer 1985: 7).

Can a theory of writing and reading move us beyond such logics and in ways that do not trivialize or literalize the tracings and inscriptions of culture? Can we conceptualize of a broader conception of 'writing' which

> gives rise to an inscription in general, whether it is literal or not and even if what it distributes in space is alien to the order of the voice: cinematography, choreography, of course, but also pictorial, musical, sculptural 'writing'[?] One might also speak of athletic writing, and with even greater certainty about military or political writing in view of the techniques that govern those domains today. All this to describe not only the system of notation secondarily connected with these activities but the essence and the content of these activities themselves.
>
> (Ulmer 1985: 9)

In the age of techno-political writing – the age of electronic media – the modern techno-sciences have fragmented eye, hand and ear and organized them hierarchically in their own institutionalized analytical frameworks. At the same time, techno-politics manipulates the media in a total onslaught which demands a different reading; one in which text and context take on very different meanings. Specifically we need a grammar that transcends, and opens up, the various specialized 'grammars' of the sciences – speaking, writing and mapping. In this sense, propaganda maps are not merely one more medium or form to be interpreted, but are in many ways an archetypical form of the age of technicity. They are exemplars of the manipulation of symbols and writing. They cannot be read without a broader grammatology than the one provided by 'map-reading skills'. It is to this broader grammatology of mapping that we now turn.

3   Situated pragmatics

Maps and mapping as social practice

> One great dark secret of the history of cartography, barely hinted at in most accounts, is that every map has to emerge from some determinable social and economic milieu. Moreover, the shape the map will take will largely be formed by the needs, tastes and technical accomplishments of that milieu. Frequently, authors write about ... [maps] ... as if they have come into existence in a sort of social and economic vacuum, as if they were the expression of some Blake-ish pure spirit.
>
> (Buisseret, *Rural Image: The Estate Plan in the Old and New*)

THE POWER OF MAPS

For over twenty-five years, Denis Wood has been provoking us to think differently and critically about maps and map use. In *The Power of Maps* Wood challenges the pretence of professional cartographers to be the objective/scientific producers, readers and interpreters of maps. Instead, he insists, we need to pay attention to the legion of map-makers and map users that is not part of the professional cadre of expert cartographers and to understand the ways in which all maps function in terms of specific sets of social interests. In particular, Wood argues, we must recognize that map-making as such (and as distinct from the general ability of mapping and way-finding) emerged historically in conjunction with capitalism and the state. The essays in *The Power of Maps* elaborate these themes and in them we see the wit and insight of one of geography's best readers of maps and the cartographic enterprise.

The central thesis of *The Power of Maps* is that maps work in the sense that they present the accumulated thought and labour of the past. As repositories of what John Berger would call 'ways of seeing', maps are both selective and interested; what Wood calls 'interested selectivity'. As we have seen in the previous chapter, through inscriptions of presences and absences maps conspire to inscribe and then to mask their own interested nature. Maps work by naturalizing themselves by reproducing a particular sign system and at the same time treating that sign system as natural and given. But, map knowledge is never naïvely given. It has to be learned and the mapping codes and skills have to be culturally reproduced so that the map is able to present us with a reality that we recognize and know. This known reality is differentiated from the reality we see, hear and feel, and this is the magic and the power of the map. The map does not let us see anything as such. Instead, it lets us see the world how others have seen it and how they want us to see it. The map opens a world to us through systems and codes of sedimented, acculturated knowledge. In this view, the map is never a representation of the real, but always a 'stretch' from the real (as that known by us in our daily life) produced by systems of abstract symbols. The map points us to a world that we might come to know provided we are willing to learn and accept – to 'buy into' – this system of symbols and icons, a coded world in which particular meaning and information is presented.

The myth of the dispassionate neutrality of the map hides the socially constructed nature of the image. In this view, the map is a transparent object that reflects like a mirror that which is real: the map is the 'mirror of nature' in which the real is represented transparently as objective, neutral and accurate. And it is this view of the map as a technical and scientific tool whose principal characteristic is its ability to represent the earth accurately, objectively and neutrally, that stands in the way of a critical theory of signs and representation. In this context, Wood (1992: 45–6) argues, the history of cartography becomes clearer: 'Transfixed, as professional cartographers so often are, by the minutia of projection and scaling, generalization and symbolization, it must be tempting to view the history of cartography as nothing more than a halting but unstoppable progress toward an unachievable Nirvana of perfect accuracy.'

David Livingstone (1992: 4–5) has referred to this type of history as Whiggish history. It is presentist in that it interprets the past from the perspective of the present. It is linear and often serves as a 'textbook chronicle' written to 'inaugurate [scientific] apprentices into the mysteries of their chosen craft' and provide students with a series of exemplary historical spectacles. Such chronicles and spectacles have the primary function of making the present state of the field fit into a plan. But such representations of disciplinary history tend to write the history of science and technology as the history of great men and women, parading these 'great figures' to plot the progress of the field from an 'unenlightened past to a glorious present'. This kind of hagiographic and progressive disciplinary history has at least three other consequences: it tends to see the history of ideas as a dredging of the past to search for precursors, looking in archival materials for what Livingstone (1992: 7) refers to as anticipations, premonitions and foreshadowings of current wisdom and practice; it tends as a result to be 'internal' to the discipline, paying little attention to the wider contexts within which ideas and practices were conceived, communicated, received and implemented (p. 9); and it writes out of the history of ideas and practice all those attempts and contributions that did not become accepted, were not followed up on, or no longer fit within the scope of what we think of as modern science and useful technology.

The Power of Maps also seeks to disrupt hagiographic historiographies and offers a different history of maps; not a dry account of the cartographic process and the mapping industry, but a lively – some might say at times outrageous – interrogation of mapping games. Mapping may mask the conditions of its own production and the contingent nature of its own constructions, but it is a serious enterprise with material consequences: 'It really is a shell game. When the aesthetic issue gets hot, switch to science and talk about accuracy, but when that bluff is called, bring on the "wet, ragged, long underwear". But as Brian Harley has testified, it's a shell game that is played for keeps' (Wood 1992: 60).

Throughout the book, Wood provides many fascinating examples to illustrate his arguments. He develops a particularly interesting discussion of the ideological nature of debate surrounding the development of the Van Sant map – the cloud-free 'photo-image' of the earth distributed by the National Geographic Society as a poster entitled *A Clear Day*. In Wood's reading, the Van Sant map is more real, more accurate, and more true to life than the earth itself, having been engineered to this stage of hyperreality; a composite of 35 million pixels were carefully selected over long periods of time and carefully pieced together to represent the earth without atmospheric 'interference'. The realistic qualities of the resulting photo-map have been widely lauded for their clarity and accuracy. Wood mercilessly unmasks this ideology of visual purity: of a message without a code, in which the impression is given of a pure reflection of something that never existed (a simulacrum). He saves some of his most vituperative language and criticism for the supporters of this map, this type of map interpretation, and the ideological project of objectivist science of which it is a part.

Against this view Wood argues that the map does not represent the terrain as such. Instead, all maps represent a particular image of the world that reveals the agency of the map-maker, usually reflecting the interested selectivity of the state. Either directly or indirectly through its support for major research organizations, the state is the major producer of maps. And so, following Brian Harley, cartography is for Wood (1992: 43) 'a form of political discourse concerned with the acquisition and main-tenance of power' and has been part of the expansionist histories of *map-making states* for generations. Even when he turns to the uses of the Van Sant photo-image to support environmental awareness, Wood's (1992: 69) critique remains direct:

> Acme of cartographic perfection though it is, the map thus emerges in the context of a map-making society *struggling with its future* to serve an interest, that of those committed to ... a certain vision of what it

means to live. You may share this vision – I do – but it serves *no* interest at all to pretend that it is the planet speaking through the disinter-
ested voice of science, instead of me, Tom Van Sant – or you.

How then are we to understand and interpret the map as a product of
situated interestedness? Wood calls for a broader concept of mapping
which recognizes the role of political and economic factors in defining and
determining which features should be characterized as 'permanent' and
hence included as features of interest on a map. These are primarily features of interest in the informational economy of the state and a capitalist
economy (boundary lines, property designators, primary routes, forest
cover, campsites, picnic sites, exposed wrecks), while features not of such
central interest are characterized as 'impermanent' and hence excluded
from the map, despite their centrality and permanence as elements of
everyday life (as Bill Bunge has shown, abandoned automobiles, green
trees and shrubs, dead trees and shrubs, rubbish, broken bottles). Wood's
point is that such questions must always be asked again and again as new
and different ideological constructions and material interests are asserted
under the guise of the natural and objective representation – the map. The
map as a product of an interested project, with specific contexts of production and underlying material interests, is always changing.

I have been tempted to say that *The Power of Maps* illustrates how the
map and mapping project can be deconstructed. Certainly this kind of map
criticism is increasingly being characterized as deconstruction. But *The
Power of Maps* is grounded in the ideas of Roland Barthes not Jacques
Derrida. At times, the two projects have much in common and both would
accept Barthes's injunction of the necessity to 'track down, in the decorative display of *what-goes-without-saying*, the ideological abuse which, in
my view, is hidden there' (quoted in Wood 1992: 76–7). *The Power of
Maps* is primarily focused on unmasking the activities of map-makers, the
reasons they made the choices they made, and the interests that those
choices and actions served; its goal is to unmask 'the ideological abuse
which ... is hidden there'. In this reading, behind every map and sign
system is an interest (or multiple interests) and a message about the role
the image was intended to play. Map criticism is about making clear the
embodied interest that drives selectivity. As a result, the interests Wood
seeks to unveil are those that have been exercised and are knowable and
expressible. The map is open to determinate and limited interpretations,
determined and limited by the selective interestedness that gave rise to
them. When these interests are unmasked, the map will become a [more]
transparent object, able to resume its true character as a 'particular view':
'Freed from being a thing to ... *look at*, it can become something ... *you
make*. The map will be enabled to work ... *for you, for us*' (Wood 1992:
183).

Since maps always and everywhere represent selective interests, *The

Power of Maps offers a series of recommendations about how maps may be developed and used to serve more democratic interests: map-making technology can be decentralized and made more accessible to the people who need it (Wood 1992: 190); map-makers can be upfront about their sources, procedures, and choices made; map critics can challenge those 'map-makers maps' and map-makers who lavish skill upon skill by way of obfuscation (Wood 1992: 240, fn 19); and the map critic can challenge the arrogance of the expert (Wood 1992: 192). Presupposed in these prescriptive recommendations is the need to push the kind of advocacy exemplified by Bill Bunge's expeditionary mapping of Detroit into a kind of genuine professionalism, whose weighty responsibility is fully recognized and acted upon (such as occurred with Bunge's decision to produce 7,367 maps to evaluate the entire range of school redistricting options that had actually been available to the Detroit School Board (Wood 1992: 186–8)).

In this sense Wood is concerned less with developing a method and theory of map interpretation than with encouraging a broader and critical awareness of map production and use, with 'constructing and reconstructing the map' (Wood 1992: 187) in ways that reveal its hidden and naturalized choices and interests. Wood destabilizes the representational understanding of maps as mirrors of nature (as naturalized or ideological), and in its place he argues for a position of advocacy, map criticism, and alternative mapping strategies – a kind of nomad cartography.[1]

Maps work by serving determinate interests, they are products of history and contribute to the construction of particular histories, they are partial representations, the interests they serve are masked, these interests are normalized and generalized through the signs and myths of map construction and use, these signs and myths themselves have emerged historically, and (as a result of this contingent construction) we can carry out forms of ideology critique and (by appropriating the act of map-making) we can begin to think of ways in which maps can be used to empower different people and serve different interests.

Cartographers have long recognized the partial and selective nature of mapping, the close association between mapping practices and military, state and commercial interests, and the openness of any tool like a map to uses fair or foul. In a similar way, Brian Harley (1989b: xx) had drawn attention to the ways in which power functions in maps through the selection and omission of content, the deployment of unmediated cultural practices and symbols, and the normalizing and universalizing of culturally specific representational forms:

> Cartography deploys its vocabulary ... so that it embodies a systematic social inequality. The distinctions of class and power are engineered, reified and legitimated in the map by means of cartographic signs. The rule seems to be 'the more powerful, the more prominent'. To those who have strength in the world shall be added the strength of the map.

In *The Power of Maps*, Wood takes this characterization to heart and asks us to think about the systemic context within which mapping functions as part of traditions and practices. One crucial context within which such map use occurs is that of the liberal state and liberal capitalism. But Wood also works with a broader conception of the power of maps in which a pragmatics of map use is being developed, and it is to this pragmatics of map use that I now turn.

'THE POWER OF MAPS': TOWARDS A SITUATIONAL PRAGMATICS OF MAP USE

> We would like science to be free of war and politics. At least, we would like to make decisions other than through compromise, drift, and uncertainty. We would like to feel that somewhere, in addition to the chaotic confusion of power relations, there are rational relations ... To this end we have created, in a single movement, politics on one side and science or technoscience on the other. The Enlightenment is about extending these clearings until they cover the world.
>
> (Latour, *The Pasteurization of France*)

In his 1993 *Cartographica* essay 'The fine line between mapping and map-making' Wood drew a sharp distinction between his own understanding of a new critical cartography and that called for by Brian Harley. The differences are epistemological, practical, and political. Wood began with the question: 'Why didn't Brian Harley write the history of cartography he wanted to?' He answers that Harley was a victim of his own idealist understanding of mapping and maps, a 'reactionary and superficial' reading that 'never penetrated to the map itself' (Wood 1993: 50).

> The problem for Harley remained the bad things people *did* with maps, and ultimately this left the maps themselves out of the picture. Insulated by an idealist conception of knowledge, Harley was never able to conceive of the map as other than a representation of reality; was never able to grasp the map as discourse function; was never able to understand that the heart of the problem wasn't the way the map was *wielded* but the map function itself. His refusal to acknowledge the map as a *function* of social being – not just as something *colored* or *shaped* by this or that social vector – prevented him from seeing that map-making was *not* a universal expression of individual existence (like something we might call mapping), but **an unusual function of specifiable social circumstances arising only within certain social structures**.
>
> (Wood 1993: 50) (Italics in original; emphasis (bold) added)

For Wood the practice of map use is not to send a message, but to bring about a change in the way another person, or group of people, see the world. It is 'out of their interaction in the social worlds they inhabit that people bring forth cultural products like maps' (p. 52), and such cultural products act to induce social, economic and political change as 'weapons in the fight for social domination'. Thus, for Wood, such a pragmatics of map use requires a more radical shift in cartographic epistemology than Harley was able to accept.

> For all the political self-consciousness that is so exciting in Harley's late papers, there is still the same stuffy quality that Harley hoped he was opening the windows on. Despite, for example, the derivation of the title of 'Victims of a Map' from the title of a collection of poems by the contemporary Palestinian Mahmud Darwish (and others) there is no sense in the paper that Harley is dealing with a general problem of contemporary relevance, his history is not living ... his victims all turn out to be native Americans who died centuries ago, they remain sealed in the past, there is even little sense of the social *co-construction* of the New World.
>
> (Wood 1993: 52)

In Wood's analysis, it was Harley's inability to shed the inherited idea of the map as a representation of the real world and his inability to accept really that the map was a social construction of reality that prevented the emergence of this new critical social cartography. He agrees with Barbara Belyea's (1992) reading that Harley remained an idealist and a British empiricist, despite his efforts to accept the discursive and deconstructive critiques of Foucault and Derrida. When Harley asked whether a normative ethics was possible or were we left with a 'slide into a cozy relativism in which cartographic values vary with different societies, generations, social groups, or individuals?' (Harley 1991: 14), he overlooked the fact that maps cannot but embody such social situations and desires. For Wood (1993: 53): 'writing is *not* captured speech, which was never *thought* put into words, behind which was *never* anything ... *real*, anything ... *true*. Nothing ... *behind* ... the map guarantees it. *Or* throws it into doubt.'

Maps are made because of the needs of particular social situations; they are made to fulfil a particular function. As a result, there cannot be a general theory of mapping and cartography, only a pragmatics of map-making and map-using. As Wood (1993: 53) argues, the map 'exists in its inscription. And it is the fine line of this inscription that differentiates something we might call mapping (but which is really just ... getting around [forms of spatial competence]) from map-making; and mapping *societies* from map-making *societies*.' This situational pragmatics of mapping focuses on the 'map's discourse function' (p. 56) asking 'not what does the map *show* or *how* does it show something, but *what does the map*

do? what does it accomplish?' (Wood 1993: 56). 'It is the inscriptive property of the artefactual map that permits it to serve the interests of the
power elites who control the map-making process (as well as those who
would contest them)' (Wood 1993: 53).

> I believe people for millions of years have emitted map, and maplike
> and protomaplike, artefacts as natural consequences of their spatial
> competence working itself out in the context of human discourse
> about the territory and what comes with it; but I also believe that most
> of these have been one shots, squibs, duds. Or they've made their
> point ... but no one noticed. In neither case did they lead to map-
> making. Not until the demands of agriculture, private property, long
> distance trade, militarism, tribute relations, and other attributes of
> redistributive economies transformed the discourse environment in
> which these firecrackers exploded was the light they emitted apparent.
> But then maps must have seemed the answers to prayers (*why hadn't
> anyone thought of them before?*).

I began this book with Gunnar Olsson's critique of cartographic reason
and with his provocative question: 'What is geography if it is not the
drawing and interpreting of a line?' Perhaps now we begin to see more
fully how important is Olsson's question. If we are ever to understand this
process of 'drawing a line' and by extension the processes and practices
of mapping, it will be useful to have some clear idea of what actually
happens when lines are drawn and maps are made. In various
ways, Harley and Wood have helped us to understand how lines are
selected, drawn and accepted within a community of users. How are some
symbol systems drawn into the domain of cartographic practice while
others are not recognized immediately as being non-cartographic or bad
cartography? How does the map get produced? Precisely how and under
what conditions do particular mapping forms and conventions arise
as standards for the social practice of mapping? And, how do these standards and practices get reproduced and normalized as 'sound cartographic
technique'?
    At stake is not merely an expansion of the self-understanding and prac-
tices of map-making and map use, nor is it merely a challenge to the traditional conceptions of 'objectivity' in mapping sciences. It is, beyond all
these, a fundamental question of how maps work in practice; a situated
pragmatics of map use that begins with the clear understanding that what
the map represents and the ways in which it represents the world are not
guaranteed by anything behind it. It is not a representation of the world,
but an inscription that does (or sometimes does not do) work in the world
(see Curry 1996). It is this that Wood points to when he says that maps are
instruments of power embedded in and reflecting the social relations and
interests that give rise to them. Here is not a functionalist reading of maps,

but a pragmatic reading of post-representational cartography – a pragmatics with political intent.

MAPS AS SOCIAL PRACTICE

> It has always seemed that if a science were not independent of politics, something would be missing and the sky would fall on our heads. To show that the sky holds up perfectly well on its own, we have to be able to prove in a particular scientific discipline that belief in the sciences, like the old belief in God, is a 'superfluous hypothesis.' We have to give evidence that 'science' and 'society' are both explained more adequately by an analysis of the relations among forces and that they become mutually inexplicable and opaque when made to stand apart.
>
> (Latour, *The Pasteurization of France*)

Because the technologies with which we live more or less work as they are supposed to, we tend not to ask why or how any particular technology or ensemble of technologies work, or why they came into being in the first place. Most of the time, most of us take them for granted.[2] We certainly tend not to ask about the design decisions, the logics and the rejected alternatives that went into the selection of particular paths to the construction of the technologies with which we work today. We probably think even less about the professional, political, economic and social contexts within which these decisions and choices were made, or about the ways in which they were put into practice. Even when a problem arises, our first response is more likely to be one that seeks a technical solution to fix the problem instead of asking about the broader context of origins, development and practice within which the technology works or doesn't work.

In this sense I am reading Wood's argument that nothing lies behind or guarantees the map as a radicalizing of the deconstructive impulse sought by Harley. Wood is correct, in my view, in recognizing the limits of Harley's actual 'deconstructions' and more successful, as a result, in destabilizing the ontological commitments of cartography to a representational epistemology. But he also takes us too quickly to a determinate (at times perhaps functionalist) reading of the power that shapes the pragmatics of map-making. His historical and institutional readings of maps are rich and provocative, but they too seek to uncover the shaper of the message and the power *behind* the map in much too literal a manner. As we move forward, I shall deepen this deconstructive turn on precisely this point, focusing more directly on the multiple and disseminated practices of mapping and map-making, and on what I hope will be a more articulated and contextual reading of the cultural politics of maps.

For the moment, we can perhaps evoke a stronger metaphor than that of archaeology, of an unveiling or uncovering, which reduces modern cartography as a social practice to a single narrative. It might be better to understand cartography as more like a series of technological, scientific, and rhetorical trails in the woods. Like animal trails in the woods, trails emerge from the discrete choices and the concrete goals of walkers and/or animal. These choices are constrained by the prior uses of the forest and the ways in which others have previously passed through them. In choosing and hence building such pathways, animals might begin to consolidate their route-seeking along a central trail (the herd trail), while others might cut off in branches, create alternative trails or strike out in different directions. There is no necessity for such trailblazing to produce 'optimum' pathways, only pathways that succeed in getting from one point to another through the woods. Nonetheless, there is every likelihood that convergence around one or more central trails will occur, with gradual consolidation of that trail over time – the accepted trail to follow; the trail becomes naturalized.[3] There is, in other words, a kind of path-dependent convergence of multiple forms around specific notions of efficiency and appropriateness. But there is also every good reason to believe that alternative paths are not only possible, but are already co-present in what appears to be a single 'standard' and dominant set of norms. In this 'ecological' path-dependent model there remain strong interests and an overwhelming presence and power of the state. But there are also minor chords, off-track paths and counterveiling tendencies that must be incorporated into an understanding of the structures of power and influence. Perhaps we can think about mapping practices in these ways, and use such 'overdetermined' notions of technical and scientific change to think how we understand the history of mapping and cartographic reason. As we shall see, this question brings us to the heart of a series of debates about contemporary uses and practices of mapping and the use of the cartographic imagination.[4]

Like Denis Wood, cultural studies of science see science in terms of social practices. But in contrast to Wood's efforts to clarify the determinate interests that produce and are served by a map, science studies expresses caution in historical explanation that fixes responsibility too simply on any particular institution or interest:

> Even if a few people still believe in the naïve view, courageously defended by epistemologists, that sets science apart from noise and disorder, others would still like to provide a rational version of scientific strategy, to offer clear-cut explanations of how it develops and why it works. They would like to attribute definite interests to the social groups that shape science, to endow them with explicit boundaries, and to reconstruct a strict chain of command going from macrostructures to the fine grain of science. Even if we have to give up our beliefs in science, some of us still wish to retain the hope that another science,

that of society and history, might explain science. Alas, as Tolstoy shows us, we do not know how to describe war and politics any better than we know how to explain science.

(Latour 1988: 6)

In this sense, science studies opens map studies to a much richer social and conceptual analysis than discourses of 'maps-as-power' have been able to do. Understanding scientific practice involves understanding how new 'machines' and disciplined human performances and the relations that accompany them are constructed and interactively stabilized (Pickering 1995: 21). This requires historicizing and contextualizing the universalizing claims of science to serve as a privileged form of objective reason within concrete geographical and historical settings, and showing how technological and scientific systems are outgrowths of human practices and decisions that are locally situated. It becomes important to trace the ways in which individuals, technological objects and institutional assemblages have functioned to naturalize one particular understanding of scientific practice. By so denaturalizing what counts as 'the history of the field', we seek to uncover paths not taken; to re-place the 'monolithic textbook chronicles' with a history enlivened by multiple actors (people, technological objects and institutional assemblages) and competing claims to truth, accuracy and use value. One task of deconstruction, then, is to write denaturalized concrete histories of multiple technological and scientific projects that on the surface appear as a unity (Haraway 1991).

As Ian Hacking (1982, 1992a 1992b) has argued, what counts as scientific reason is not constant, but changes throughout history. Different styles of reasoning become accepted as dominant and 'most reasonable' in different time periods. Reasoning based on statistics, for example, does not become 'reasonable' until the mid to late seventeenth century. In describing such grounded and bounded rationality Latour (1988: 15) urges us to 'follow scientists and engineers around' – to track them through time and space, and to map out the interconnections of the institutional, cultural, and professional strata they create and work with and within, and that others generate around them. Pickering (1995: 221) suggests that we 'explore the ways in which particular machines, disciplines, styles of reasoning, conceptual systems, bodies of knowledge, social actors of different scales, the inside and the outside of the laboratory, and so forth, have been aligned at particular times and in particular places.' In all such studies, the methodological injunctions guiding such work are: refuse analyses that become 'sociologising reductions' and that reduce science to its 'social conditions'; reject analyses that provide satisfactory analysis only of the applications of a science, but fail to address its technical content; and avoid all recourse to the 'folklore of the people studied (terms such as "proof", "efficacy", "demonstration", "reality", and "revolution")' (Latour 1988: 9). The challenge for such disciplinary histories is to understand at one and the same time the *content* of science and its *context* in terms of specific practices, actors and institutions.

In *Pandora's Hope* Latour illustrates one such science study through an account of the way in which citations are circulated and consolidated as facts in science. He describes a situation in which a geographer (who is a geomorphologist), a botanist, a zoologist and an anthropologist (Latour himself) carry out fieldwork in the Amazon rainforest. From 'raw field' to 'completed categories' Latour describes the various ways in which what he calls the 'circulation of citations' begins to build up 'scientific' categories, to render literally and map the 'raw field' as a map of discrete and relational objects for scientific investigation. In this process, labelling, annotation, categorizing and mapping literally circulate among the participants, at first through their notebooks and later among their laboratories. Circulation generates eddies of attention, commitments to specific abstractions, and fixations on one particular rendition of categories or mappings. It is to these processes and practices by which cartographers produce the real as a historical and social process of circulating and adjudicating citations and inscriptions that we now turn.

Part III

The over-coded world

A genealogy of modern mapping

an administrative and political space was articulated upon a therapeutic space; it tended to individualize bodies, diseases, symptoms, lives, and deaths; it constituted a real table of juxtaposed and carefully distinct singularities. Out of discipline, a medically useful space was born.

(Foucault, *Discipline and Punish*)

In Southeast Asia, the second half of the nineteenth century was the golden age of military surveyors – colonial and, a little later, Thai. They were on the march to put space under the same surveillance which the census-makers were trying to impose on persons. Triangulation by triangulation, war by war, treaty by treaty, the alignment of map and power proceeded.

(Anderson, *Imagined Communities: Reflections on the Origins and Spread of Nationalism*)

For myself all I ever wanted was that perfectly pitched telescopic perch from which to view the earth; a gradual approach to a position near its face where I could just begin to hear the low hubbub – the sum of all creatures' voices. I sought to treasure that moment before we are too close and begin to distinguish locales and dialects, for we cannot see or hear the whole properly unless we are a respectable distance. The whole earth is my treasure; my beautiful package.

(Wortzel, 'Globe theater archives: a blue planet discourse')

Our modern civilization is to a great extent based on ... identities and differences ... [D]iscourses on the countries and peoples outside Europe, particularly 'the Oriental', have been a part of the European's power relations that constitute the presence of 'the Other' in order to confirm the identification and, more often than not, the superiority of the European metropolis itself, rather than being the documentation of what 'the Oriental' actually is.

(Thongchai, *Siam Mapped: A History of the Geo-Body of a Nation*)

4 The cartographic gaze, global visions and modalities of visual culture

> The point is not that social life is guaranteed by some shared visual culture, neither is it that visual ideologies are imposed on individuals. Rather, it is that *social change is at once a change in the regime of representation...*
> (Fyfe and Law, *Picturing Power: Visual Depiction and Social Relations*)

> I employ the word 'picturing' instead of the usual 'picture' to refer to my object of study. I have elected to use the verbal form of the noun for essentially three reasons: it calls attention to the *making* of images rather than to the finished product; it emphasizes the inseparability of maker, picture, and what is pictured; and it allows us to broaden the scope of what we study since mirrors, maps, and, as in this chapter, eyes also can take their place alongside of art as forms of picturing so understood.
> (Alpers, *The Art of Describing: Dutch Art in the Seventeenth Century*)

INTRODUCTION

To ask what a map is and what it means to map the world in a particular way are, in part, technical questions about the ways in which information and ideas are represented from one domain of reality into another. It is to ask about those 'acts of visualizing, conceptualizing, recording, representing and creating spaces graphically – in short, acts of *mapping*' (Cosgrove 1999: 1). That is, mapping is about the transfer of information from one form of presentation into a re-presentation of that information – be it empirical information about the earth, systems of belief about a society, symbolic, mythic or dream forms dependent upon a depth hermeneutic for their re-presentation, or formal mathematical relations of translation or transposition.

As we have seen, information is itself not merely given to us naïvely, but is itself a product of norms, standards, values and interests. While some of these rules and standards (such as those relating to map projection) change slowly, if at all, others are more plastic and depend on specific social conditions. As a result 'mapping' remains a slippery concept. Few concepts exercised Peter Gould more than 'mapping', and one of the

abiding lessons of his teaching was the importance of clarifying the different meanings of mapping through the use of different kinds of mathematics. In this sense, his commitment to *mathesis* as the art of mapping social
and spatial relations mathematically can, I think, be seen as an attempt to
dislodge the hold of representational thought on geography and cartography. In its place n-dimensional mappings of one kind or another can be
seen to be replacing representational logics with a much more active sense
of the role of the cartographer in choosing and shaping the forms of
abstraction and space with which we deal (Figure 4.1).

    To ask about the map and the mapping process is, then, also to ask
about the systems of social beliefs and practices that give rise to the
mapping project, the rules that govern the translation and transposition,
and the specific work done by a particular mapping or map. It is, that is, to
ask about the systems of belief within which particular procedures of
representation are accepted and judgements are asserted about what constitutes valid or effective forms of representation. To ask what a map is
and what it means 'to map' is also to ask about the epistemological and
ontological structure of the world in which we live and map. These epistemologies and ontologies are, of course, not simple things. They are
complex assemblages of disparate, contradictory, and overlapping beliefs
about the world, and their differences often give rise to quite different
understandings of the map and the mapping process. To ask what a map is
and what it means to map, therefore, is to ask: in what world are you

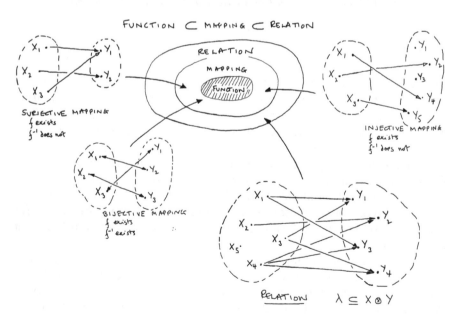

Figure 4.1 'Multi-dimensional mappings': Peter Gould's surjective, bijective and
        injective mappings (with permission, Jo Gould)

mapping, with what belief systems, by which rules, and for what purposes? It is to these issues that we now turn. In particular, we turn to one particular set of beliefs and assemblages: that of the priority given in western thought to vision as a privileged form of social practice, and through it to a particular and singular conception of representation.

Much discussion of maps and mapping focuses on the ways in which historical transformations in social life and thought have influenced mapping techniques and map use. But for the moment, I want to turn this around and ask: how was it that social life and thought were affected by forms of cartographic reasoning? How did cartography function as a metaphor and model for thought and practice more generally? How are we to understand Farinelli's claim that western thought is cartographic, or Francis Bacon's (1605) account of his own science being a kind of cartographic reason: 'Thus have I made as it were a small globe of the intellectual world, as truly and faithfully as I could discover' (quoted in E.O. Wilson 1998: n.p.)?

The idea that cartography served metaphorically and conceptually to shape social thought more generally has, of late, begun to transform the history of cartography. In the place of a disciplinary history of technical change linked to practical pursuits, we are beginning to understand the much more complex relations between cartography and the broader arenas of social thought and action. Indeed cartographic reason seems to have been so powerful a force in the sixteenth and seventeenth centuries that it came to signify the most important forms of reason. *To map was to think.*[1] With this change in our thinking about cartographic reason comes a much stronger geographical imagination.

At the heart of this social imaginary of modernity was the privileging of vision and sight. If seeing is a cultural experience, as John Berger has shown, there are important differences in 'ways of seeing'. For our present purposes, I discuss two ways in which these 'ways of seeing' can help us to rethink our understanding of maps and mapping. The first has to do with the priority of cartographic reason in western thought and life. The second has to do with the geographies of cartographic reason and ways of seeing. In this section, I deal with three aspects of the cartographic gaze, three ways in which vision was prioritized as the primary mode of cartographic 'visualization', and the effects this had on the form of maps. I will deal in turn with the role of perspective, the importance of projection and the issue of the construction of accuracy. My goal is first to clarify the ways in which the cartographic gaze has been understood, and second to show how perspectivalism and projectionism illustrate two geographically specific forms of vision and representation – what Svetlana Alpers has called the Southern and Northern schools – in which the techniques of observation, painting and mapping came into being quite differently and for different reasons. The first points to an intellectual history of geographic representation, the second to a geography of representations. The history of accuracy – rendered differently in each school – is one in which differences have collapsed around a cartographic anxiety (a direct analogy to the Cartesian anxiety we discussed previously). I turn then to bricolage and montage to provide different metaphors and models for understanding actually existing cartographic practices.

ENLIGHTENMENT REASON, MAPPING AND THE DREAM OF THE FUTURE

In his Paris Arcades project (carried out between 1927 and 1940, the year of his death), Walter Benjamin suggested that '[e]very epoch dreams the one that follows it' as the *dream* form of the future, not its reality (Buck-Morss 1989: 116). Perhaps one of the abiding dreams of modern science has been to map the globe in its totality; to map 'everything' and to map it as a unity. Global mapping and 'globe hopping' were a part of the sixteenth- and seventeenth-century project of European exploration and science, a nineteenth-century project of territorial acquisition and ethnographic taxonomy, and a twentieth-century project of imperial reach (see Driver 2001: 204). As a consequence, images of the globe abounded in popular media.

We tend to think of the release of the 'whole earth' images taken from Apollo 8 (Earthrise, 1968) and Apollo 17 (Whole earth, 1972) as a major stimulus to a global vision of an interconnected, fragile earth, *Gaia*. Certainly *Gaia* stimulated enormous interest in global concerns and consolidated a western trope of global unity and concern about western economic and environmental power (Figure 4.2), becoming as it did the icon for the *Whole Earth Catalog* and – for a few years at least – a standard symbol for alternative thinking about nature and society. But such global images have a much longer heritage in western popular thought. From atlases to national exhibitions to commercial advertising for imperial products, the globe has circulated as an image, icon and trademark for science, technology, imperial power and commercial vitality since the Renaissance.

More recently, such global mappings have become generalized – almost universalized – to all domains of life, especially those involving technology, speed and power. Technical changes have made the cartographer's dreams of mapping the earth as a whole even more of a reality. They have transformed the ways in which we map the earth and society, and the ways in which we think about both. Images of 'whole earth', representations of relationships that transcend local, regional or national identities but which embed each in wider grids and networks of interaction and meaning, new notions of community that transcend parochialism of place or locality, and new mediations of self and society and self and nature have all become realities with new computerized mapping and imaging technologies. Their iconic status has been so thoroughly entrenched that the whole earth

Figure 4.2 'Whole earth', Apollo 17 (NASA)

image now can mobilized to serve needs as diverse as the exercise of war, the location of commercial and service outlets, local government planning and regional service provision, corporate marketing strategies, and – of course – science itself. In all these domains of life, computer cartography, geographical information systems, and remotely sensed data of one sort or another (often in combination) have come to play ever more important roles in mapping and representing the earth:

> From 'comparative planetary geomorphologist' to 'nanotopographers' – map-making has seized the cultures of science in ways that are redrawing the maps of the world ... The last time such wanton map-making seized the culture was not long after the establishment of the Gutenberg press. Then, a *Novus Mundus* had been discovered, and the map of the world had to be literally redrawn. The current upheaval in cartography hints that a shift of comparable magnitude is under way.
> (Hitt 1995: 26–7)

One reason for this wanton map-making must be seen to lie in the importance of visual imagery in western thought. Theodor Adorno (1973: 139–40) captured this interest in terms of a metaphysics of the gaze suggesting that: 'Except among heretics, all western metaphysics has been peephole metaphysics ... As through the crenels of a parapet, the subject gazes upon a black sky in which the star of the idea, or of Being, is said to rise.' One consequence of this way of making the earth visible is that nature, earth and space are rendered as a resource, as a source of information and value, in which all information will be available in one place at one time. Such a universalism and transparency grounds the scientific world-view and its dream of utopian communities of openness, reason and democracy.

But what precisely were the technological and material conditions that produced this visual metaphysics and scopic regime, and what were the ideational and ideological contexts within which the 'value' of the gaze emerged? Perhaps more interesting for our present purposes, how did this modern metaphysics of vision emerge and in what ways was it related to the practices of cartography? In short, what is the cartographic gaze and what roles does it play in western thought? It is to these questions that we now turn our attention.

THE CARTOGRAPHIC GAZE

By the term 'cartographic gaze' I mean something quite specific and technical. As we will see later, there are many potential forms of cartographic representation and cartographic imaginings and I do not want in any way to foreclose on the limits of these at this stage. But in using the 'cartographic gaze', I refer to the particular constellation of ways of seeing with its particular practices and institutions of mapping that emerged in the modern era. The 'cartographic gaze' thus has several distinct characteristics each typified in the cover illustration for this book. It assumes what Adorno called a 'peephole' metaphysics, an observer epistemology, and a Cartesian commitment to vision as the privileged source of 'direct' information about the world. It presupposes what Martin Heidegger called 'world space', a parametric manifold within which nature and society can be thematized in terms of their spatial relations. It has prioritized mathematical forms of abstraction over other forms of abstraction in this process of thematization. It has come to see itself as a technical–scientific practice of representing (mirroring) nature. It has accepted a universalist logic, underpinned by commitments to particular forms of parametric space, geometry and scale. It is, above all, a controlling gaze rendering the broad swathes of worldly complexity and enormity in miniature form for a discrete purpose. And, as the history of mapping demonstrates, the ordering principles of the cartographic gaze have always had political intent and/or consequences. The cartographic gaze is dominated by a commitment to modelling a God's-eye view, what Donna Haraway (1991) called the 'God-trick'. This transcendental positioning is both the view from above, an elevated two-point perspective bird's-eye-view, *and* an all seeing eye that views everywhere at the same time.

Perhaps one of the most difficult lessons for anyone to learn is the way in which their own worlds are geographically coded; to understand the relationship between the visible and the invisible, the proximate and the distant, and to recognize the complex folds of past and present that constitute place and experience as we know it. The modern world-view has been so lodged upon naturalist notions of experience that people are very reluctant to question their belief in the stability and reality of their visual world. How could it be otherwise? Isn't the world clearly apparent to us? Isn't sight the purest of all our senses, the one least affected by social values and cultural practices? Well, it turns out that it isn't so clear or so stable.

The technologies of vision are complex and have been developed with great difficulty over hundreds of years. They encompass not only the material technologies of the lens, telescope, theodolite, microscope and the myriad other apparatuses that have emerged in conjunction with the technologies of vision (Figure 4.3), but also the social 'technologies' by which we come to know and order the world visually, the ways it is inscribed and made 'obvious' and stable to us. These technologies include the difficult and hard-won practices of perspectival drawing, mapping and picturing, technologies that now seem so natural and normal. As the photograph *The Surveyor* reminds us so clearly, such practices and technologies of vision also depended on being able to range far and wide across space, a ranging and surveillance that in turn tied the social practices of mapping to those of territorial policing in direct and important ways.

Elsewhere I have written about the implications of this prioritizing of vision and of the politics of surveillance to which it gave rise (Pickles 1991, 1995). Martin Heidegger's apocalyptic warnings about the destiny of modern people to capture Nature as resource – to render the world-as-picture – become clearer in the light of this broader peephole metaphysics and particularly in the ways in which it valorized the technologies that enhance vision.

At the beginning of the modern age René Descartes (1965: 65) laid out the ground plan for this love affair with vision and mapping: 'All the management of our lives depends on the senses, and since that of sight is the most comprehensive and the noblest of these, there is no doubt that the inventions which serve to augment its power are among the most useful that there can be.' In a related vein, Chamberlain (2001: 318) has recently suggested that, as a result of the privileging of vision in the Renaissance, a shift occurred from 'reading the world as an intelligible text (the book of nature) to looking at it as an observable object (a secular autonomisation of the visual) . . .' an emphasis that enabled a 'new world' to be seen and made. In this sense, Harvey (2000: 220) suggests, cartography emerged as institutionalized practice,

> locating, identifying and bounding phenomena and thereby situating events, processes and things within a coherent spatial frame. It

Figure 4.3 The Surveyor: Robert Harvey (with permission of the Nebraska State
    Historical Society Photograph Collections, Lincoln)

imposes spatial order on phenomena. In its contemporary manifestation, it depends heavily upon a Cartesian logic in which *res extensa* are
presumed to be quite separate from the realms of mind and thought
and capable of full depiction within some set of coordinates (a grid or
graticule).

Denis Cosgrove (1994) has also focused on one of these techniques of
the observer, Cartesian perspectivalism, to show how it functioned as the

dominant (and dominating) mode of modern vision. In his essay 'Contested global visions' Cosgrove (1994: 271) argued that the view from the Apollo spaceship ('an Apollonian perspective') is 'implicit in Ptolemaic cartography's positioning of the observer at sufficient distance from the spherical Earth. The fifteenth-century rediscovery of this mode of terrestrial mapping marks the beginning of European Modernity.' This modernity is one which privileged a particular form of seeing (distanced, objective and penetrating), predicated on an epistemology and politics of mastery and control of earth, nature and subjects.

These commitments to abstraction, reason, science, representation, universalism and transcendence demonstrate for Gillian Rose (1995: 761–81) the role of this dominant visual system in the gendering of modernity, reflecting a deep commitment to masculinism, mastery and control. Such modernist notions of reflection and self-reflection presume a 'mirror' metaphysics, which itself presupposes a particular kind of space and subject:

> discussions of self-reflection assume a self from which the self can be distant in order to see itself, and this requires a distancing, separating, visualised space. Specific articulations of subjectivity mobilise specific organisations of space, then, and such modes of spatializing the self are so deeply bound into ways of understanding the world. That is, the spatialities of subjectivities and the spatialities through which the material world is represented mediate one another.
>
> (Rose 1995: 762)[2]

Rose's essay is challenging on several fronts, but for our present purposes I want to focus on the ways in which she draws from Irigaray a correspondence between a certain kind of modern knowledge (detached, objective, distanced, scientific), a particular kind of surveillance (predicated on a distanced gaze, a detached observer epistemology as Helen Couclelis (1988) once called it), and a certain kind of space (one in which analysis, calculating, measuring and surveying render the world as an ordered, perspectival space – 'the perspectival space of the masculine subject') (Rose 1995: 763). Any challenge to phallocentrism 'must also challenge its visual and spatial organisation' and seek to create new conceptions of space–time, places, inhabiting and identity (Rose 1995: 763–4). The Cartesian privileging of sight thus lies at the heart of the power of phallocentrism, providing – as we saw earlier – the foundational metaphor for the objectifying distance between self and other. In this sense the gendering of identity is transformed into a spatial issue.

GEO-SCOPIC REGIMES

In discussing the enframing of which Martin Heidegger wrote, I have already suggested that the 'world as picture' referred to a very distinct understanding of representation. In this understanding, earth and society were represented as standing reserve – a resource to be used. It was rendered as world–space and projected as *ta mathemata*, as a mathematical manifold. The projection of the world as mathematical was, for Heidegger, one of the fundamental ways in which modern metaphysics understands itself and the foundation for the modern sciences and for technology as we know them. Map projections are examples of this mathematized and abstracted world space. This is no less true in the world of vision. The invention of a mathematically based system of vision – perspective – during the early years of the fifteenth century and the emergence of the Dutch mapping impulse through the deployment of the mathematical projection doubly structures the emerging mathematical world-view described by Heidegger and the development of the mathematical world spaces of map projections.

Contra Rose, however, there were (and are) very important technical differences between perspectivalism and projectionism, and there are important epistemological reasons for seeing them as distinct, albeit overlapping visual systems. They emerged as distinct visual systems of representation in different parts of Europe at different times to deal with distinct and different needs and to serve distinct and different interests. Jay argues that two scopic or visual regimes in particular came to dominate European arts and sciences. One was Cartesian perspectivalism, in which the techniques of the observer are given priority and the mathematical uniformity of the Renaissance perspective grid becomes the organizing principle for all visual representation and seeing.

In *The Renaissance Rediscovery of Linear Perspective*, Samuel Edgerton (1975: 3–4) provided an interesting account of the history of cartography in terms of this rediscovery of linear perspective:

> More than five centuries ago, a diminutive Florentine artisan in his late forties conducted a modest 'experiment' near a doorway in a cobbled cathedral piazza. Modest? It marked an event which ultimately was to change the modes, if not the course, of western history ...
>
> If we are to point to a crystallizing moment for the postmature birth of geometric linear perspective, it has to be this nameless day in Filippo Brunelleschi's out-of-doors demonstration in 1425. ... In modern-day terms, his Baptistery-view experiment could be called 'the production of the vanishing point.' And to the fifteenth-century Florentine citizen, the result was almost magical...
>
> ... Directly or indirectly, it had implications which extended irre-

versibly to the entire future of western art – and to science and technology from Copernicus to Einstein.

In terms of his own times, Brunelleschi's pocket perspective demonstration was the crucial event ... in several decades of fitful, teasing progress in optics by the artists, mathematicians, and cartographers of that ... city.

The other was associated with the descriptive school emerging around Dutch seventeenth-century art. Here the aim was not to impose a particular technology and mathematics to represent three-dimensional vision in two dimensions (as with perspective) but to develop techniques of representing the world in two dimensions. Svetlana Alpers called this '*the mapping impulse*' and pointed to its exemplification in Dutch painting and cartography.

In these two distinct 'scopic regimes' we can see different purposes, positionalities and powers at work (Figure 4.4). The positioned viewer, the frame and the definition of the picture are quite distinct in each. The convention of perspective was

unique to European art and ... was first established in the early Renaissance, centers everything on the eye of the beholder. It is like a beam from a lighthouse – only instead of light travelling outward, appearances travel in. The conventions called those appearances *reality*. Perspective converges on to the eye as to the vanishing point of infinity. The visible world is arranged for the spectator as the universe was once thought to be arranged for God.

(Berger quoted in Jay 1994: 54)

Perspectivalism presupposed specific notions of space and what was visible in the perceptual field:

a homogeneous, regularly ordered space, there to be duplicated by the extension of a gridlike network of coordinates.... The result was a

Figure 4.4 Artist Drawing a Reclining Woman, 1538, by Albrecht Dürer. This image was one of several drawn to demonstrate the techniques and apparatuses for correct perspective drawing

> theatricalized 'scenographic' space ... uniform, infinite, isotropic space
> that differentiated the dominant world-view from its predecessors, a
> notion of space congenial not only to modern science, but also, it has
> been widely argued, to the emerging economic system we call capitalism.
>
> (Jay 1994: 57)

Central to so many such technologies of visual perspective was the gendered gaze to which Rose refers.

The new seventeenth-century descriptive art – the mapping impulse – was, by contrast, very much more concerned with how we know the world and how what constitutes 'public knowledge' can be represented in ways we all can understand. It was concerned with what constitutes facts and information, and how these are determined at particular places and times. And it was about the ways in which the framing of 'universal knowledge' is articulated. As Richard (1996: 71) argues:

> symbolization operates as an image of totality, establishing a fixed
> point which permits the measured evaluation of relationships of prox-
> imity and distance that either draw together or separate all other
> points distributed in space: 'Each historical period or cultural tradition
> selects a fixed point which functions as the centre of its current maps,
> a physical symbolic space to which a privileged position is attributed
> and from which all other spaces are distributed in an organized
> manner.'

Thus, alongside the phallocentric gaze and representational epistemology of perspectivialism is another mapping impulse that does not presuppose the privileged position of the single elevated view, but instead evokes a quite different relation between viewer and viewed.

CARTOGRAPHIC *BRICOLAGE* AND MONTAGE

In Italian perspective painting and in the descriptive painting of the Dutch school two different and distinct scopic regimes emerged in Europe at about the same time, each representing very different positionings of the subject, each with distinctly different claims on the observer. These 'techniques of the observer' (Crary 1995) have important implications for the ways in which we understand 'the power of maps'. They begin the process of unpacking Adorno's more generic 'western peephole metaphysics' and assessing the extent to which diverse scopic regime have geographies that have shaped our present understanding of landscape, view and map. Jay has shown how such distinct and different scopic (or visual) regimes emerged at different places and times, with their own distinct geographies and histories, and Alpers shows how a differentiated cartographic impulse was at work in the visual (or scopic) regimes in the north and the south.

In ways reminiscent of Jay's and Alpers' arguments that we need to abandon totalizing accounts that see in western metaphysics a single epistemology of sight, Alan Pred and Michael Watts (1992) have suggested that we might usefully stop thinking in terms of such universal categories as 'European Modernity' and a single modernity, and instead we might pluralize modernities and capitalisms. In a parallel way, Rose turns away from the 'overly general' conceptual categories of phallocentrism and argues instead that, in order to challenge the ways in which the phallocentric subject has been complicit with the organizing of the visual and the spatial in the modern world, it is necessary 'to argue that this position is not monolithic but is composed of fractured, contradictory, mobile, and diverse masculinities' (Rose 1995: 765). Without such an interrogation, feminist theorizations of space and subjectivity reflect 'the self-image of the same' (Rose 1995: 776). We need, that is, reworkings of visualized spatialities and subjectivities that do not presuppose the phallocentric gaze, but which subvert it, which provide the possibility for 'repetitive looking' that does not reflect the same. Rose continues these reworkings through Lacanian psycho-analysis, evoking a series of metaphorical images and corresponding epistemologies: the cracked mirror, the shattered mirror, and the picking over of the shards. I shall return to these images in Chapter 10, but here I want to step back to Rose's adequation of phallocentrism with perspectivalism and the gaze. In doing so, I want to both further problematize the representationalism at the heart of modern thought and the mirror image that reproduces phallocentrism.

In coming to grips with the massive transformations of post-communist societies in the 1990s, Stark and Bruszt (1998) have suggested that we must think of the ways in which transition involves the 'building on and with the ruins of communism'. That is, that social change is more like a process of appropriation, grafting on, and reworking with already available resources, capacities and social relations than it is a break with the past. I find this image of 'building on and with the ruins' to be a particularly suitable metaphor for how we might think about mapping practices; how we might in fact see mapping as a social practice with embodied social relations, rather than – as with most traditional histories – a discrete professional or technical activity. And how mapping as a social practice is also a historical process of accretion and reworking: a process of sequent occupance, a palimpsest of epistemological commitments and technical apparatuses and approaches. It is this assemblage that I refer to by the term 'bricolage'. To illustrate this here, I will draw on the notion of cartographic *bricolage* developed by Frank Lestringant (1994) to account for the emergence of mapping practices in the Age of Discovery.

Lestringant's (1994: 108) discussion of *bricolage* in the mapping practices of the sixteenth-century cartographer and cosmologist Andre Thevet describes how sixteenth-century mapping functioned through practices of borrowing, grafting and building on prior forms and practices. According to Thevet, the 'cosmographer's art necessarily involved a recourse to disparate materials, often of humble extraction ["the road maps that are in the hands of the people" according to Thevet's critic Jacques-Auguste de Thou], and left the field open to the inventive genius of a manipulator'. The first task was one of *montage*: the grafting of these fragments of empirical information (often without scale or standard projection) onto the theoretical framework, and the articulating of disparate pieces of diverse origins onto the preconstructed whole of a *mappa mundi*.[3] Often this required *collage*, the combination of two or more distinct modes of construction: 'By a collage, the cartographer juxtaposed the space of the portalan, with its canvas defining areas of winds, with ... a graduated double scale of latitude and longitude which, properly speaking, arose from the system of geographical projection' (Lestringant 1994: 111). Maps were produced of necessity in these ways and, as a result, bore the traces of past mapping practices, local systems of representation and internally contradictory forms. In modern terms, the cartographer's responsibility was one of articulating and 'smoothing out' the differences, but the map was in fact a *bricolage*. 'In effecting such a *bricolage*, the cosmographer forced together a practical cartography based on the lore of sea-going mariners, and a more theoretical cartography that subordinated the givens of experience to a rigorous method of geometrical construction' (Lestringant 1994: 112).

Cartographic *bricolage* illustrates a fundamental principle of all mapping, that:

> any given map was never established on entirely fresh ground, but always inherited from previous maps a not inconsiderable – even a preponderant – share of its information ... The map did not reveal the state of the world at a given moment, but a mosaic of data whose chronology might extend over several centuries, the whole being assembled in a floating space. These driftings, at the same time spatial and temporal, conferred a dynamism and a prospective value on the map. On it were depicted not only lands actually known, but also those remaining to be discovered ... For cosmography had a horror of the void.
>
> (Lestringant 1994: 113)

That is, the map was a construction that always drew upon disparate information sources, patched together mappings from a variety of sources, built upon cartographic techniques and taken-for-granted practices, and thus contained within it traces of these legacies. Moreover, the cartographer interpolated between the 'data points' to fill in the void. In sixteenth-century cartographic practice of *bricolage* we have some basic lessons for all cartography, and – as we shall see later – an opening to a post-structuralist and deconstructive cartography (see Chapter 10).

If *bricolage* can serve us not only to describe the origins of modern mapping practices but also as a general metaphor for all mapping practices, it also poses a challenge to modernity and linear histories of development, progress or evolution in techniques of mapping and in the flux of representational styles. Lestringant (1994: 131) himself interprets these practices in more limited terms:

> When what might be called the 'age of cosmography' came to an end, the link was undone between the lowly practical know-how of professional sailors and the refined science of the learned. The possibility of those rudimentary montages of heterogeneous data, those incessant short-circuits between distinct languages, images and sciences by which Renaissance science came to resemble a disconcerting *bricolage*, then vanished. Its art of using up the left-overs of a beleaguered and abused ancient knowledge by mixing in the most incongruous and insolent naiveties was possessed by Thevet in the highest degree.

I prefer to read these practices in more general terms in which history itself can be seen as a form of *bricolage* – as Michael Mann (1986) suggested 'always messier than our theories' – which requires a genealogical tracing of linkages and influences. In this sense, genealogy is always an interweaving of multiple related and disconnected practices, events, discourses and institutional settings: contingent, contextual and co-present. In the face of such utter contingency (as Denis Wood says, nothing guarantees the map), traditional theories and histories of maps and mapping have tended to either reduce maps to one or another interest – the progressive evolution of representation of the earth, the tools of power, or the material form of a universal 'instinct' or 'drive.' We should be wary of similarly reducing the map to a single narrative and giving it a single history.

REPRESENTING THE 'REAL'

Perspectivalism and projectionism were extremely powerful and widely used techniques of visual representation and mapping, and neither emerged nor remained distinct one from the other. I shall call this the 'cartographic paradox', a form of aporia in which two distinct scopic regimes emerged, each gradually draw upon the resources of the other, and yet in their merger continued to reproduce their contradictory differences. Because this aporia has been collapsed in the history of cartographic thought, important questions of power in maps and the possibilities for a different reading of the power of maps have not yet been recognized. In particular, the montage origins of all mappings and the plurality of mapping systems represented in any single map have (with the exception of works like those by Jay, Alpers, Conley and Lestringant) gone unnoticed ... except, of course, among practising cartographers as they exercise daily their skills of transmuting fragments into new wholes.

How do we understand this process of transmuting fragments into the 'Real'? How are the multiplicities of history, geography and visual regime rendered as coherent unities for use? How, in our world, did the real come to be seen as something that could be represented within a particular framework: the framework of abstract, parametric or non-parametric spaces, gridded, mathematized and projected. That is, how did the world come to be represented as *ta mathemata* (Heidegger 1982)? How did a new cartographic impulse (and with it a new set of technologies and practices) come into being whose initial stimulus seems to have been Ptolemy, but whose flourishing seems to have occurred between the fifteenth and seventeenth centuries (Conley 1996)?

One way of thinking about this question is through the work of Bruno Latour. In *We Have Never Been Modern* Latour (1993: 27) explains the fundamental implications of the parallel and related projects of Hobbes and Boyle.

> they are inventing our modern world, a world in which the representation of things through the intermediary of the laboratory is forever dissociated from the representation of citizens through the intermediary of the social contract. So it is not at all by oversight that political philosophers have ignored Hobbes's science, and historians of science have ignored Boyle's positions on the politics of science. All of them had to 'see double' from Hobbes's and Boyle's day on, and not establish direct relations between the representation of nonhumans and the representation of humans, between the artificiality of facts and the artificiality of the Body Politic.

As Latour (1993: 27) goes on to argue, '[t]he link between epistemology and social order now takes on a completely new meaning'. Elected leaders are presumed to speak for those they 'represent', while scientists 'represent' the facts, and the word 'representation' begins to take on two different meanings.

The challenge to the correspondence theories of truth and representation which underpin much modern science are here located in a broader enframing of modernity – a produced division between science and politics whose duality continues to delimit what is possible today, but whose boundaries and assumptions have been under serious challenge for two decades. It is, I think, precisely this produced and highly problematic dichotomy that founds a particular notion of representative democracy and empiricist science. It is a dichotomy that led Brian Harley to seek to 'deconstruct' the hidden social and political agendas of maps – the scient-

ific representations of nature and the object-world – and to seek to demonstrate how, in each representation is contained a series of political
decisions about absence and presence, about what counts and what does
not, and about how the world is to be demarcated; boundaries which
spring from and reflect the interests of power. Deconstructing the map is,
then, a project of putting into question the fundamental enframing of
modernity, it is a postmodern turn whose goal is to relocate the map and
the history of cartography in the context of a deepening of the project of
democracy, that is, the project of determining who speaks for whom, about
what, and with what authority.

The 'real' is always produced in terms of a particular economy of
capacities created through social, economic, political and technical
assemblages. It is the concrete assemblage of social, economic, political
and technical arrangements that call forth certain ways of acting and
seeing, and stimulate the construction of the very capacities themselves.
Deleuze and Guattari (1983) articulate this production of desire in terms
of machinic assemblages; the combined and differentiated processes of
deterritorialization–reterritorialization, smooth and striated space, and
the corresponding models of spaces through which each is balanced
differently. The coding of the socius is at one and the same time a
deterritorializing of prior social spaces and the reterritorializing of a new
social space. At the heart of the territorializing practices of modernity is a
cartographic impulse, comprising mapping practices that contribute to and
draw upon the coding of nature, space and social life in terms of metric
and parametric models of space and the emergence of capitalism and the
state. We can find this political economy of coding in both the map itself
and in the various visual and representational systems on which it
depends. It is to this political economy of coding that we now turn.

5 Cadastres and capitalisms
The emergence of a new map consciousness

Like other technological systems, cartography is also strongly and inevitably ideological: it involves not merely the drawing of maps but the making of worlds. Maps are not just colorings in of preset outlines or simple depictions of portions of the physical universe. Maps present entire world views, with all that phrase implies in terms of philosophical or scientific outlook, theological import, political influence, aesthetic perspective, and artistic choice. The multifarious worlds cartographers draw are far more than merely passive reflectors of particular cultural circumstances or idiosyncratic renderings of some otherwise objective reality; rather, maps are among the most powerful statements of belief in the worlds that they help to create. They are tools, to be sure, but they are inscriptive tools that allow as well as necessitate perspective; they are tools without which we cannot read and without which we cannot see.

(Tomasch, '*Mappae mundi* and "The Knight's Tale"')

MAPPING AND MAP READING

We have seen in previous chapters that the mapping impulse (as a technology for representing the real) emerged under specific conditions in different places and at different times. This historical geography of mapping has only recently begun to be unpacked, but it is clear that – as Alpers (1983) has shown – there were important differences between regional schools of imagining space and place. We have also seen how these distinct, and at times, competing scopic and mapping regimes were highly contested, especially in so far as they were part of broader political economies and intellectual traditions. From T-and-O maps to portolan charts to national cadastres, maps have been important elements of broader political economies and social formations. Not only have maps served as elements of a representational economy – what Latour called 'the modern settlement' – but, as Tom Conley has shown, they have served as the metaphysical and metaphorical basis for a broader social imagination; the world was literally and figuratively structured based on readings and interpretations of maps. What is becoming clear in all of these accounts is the way in which mapping, even as it claimed to represent the world, produced it. And it is to this process of 'world production' that I now turn.

I have always been fascinated by the ways in which maps make palpable something without existence. This may seem to be a strange claim for a practice that has always prided itself on its ability to represent accurately and faithfully that which is real. But it seems to me that the productive and fictive character of maps is precisely what is at stake when we ask, how do cartographers render the world in map form, how do maps reproduce worlds, and how can we learn to understand the worlds maps contain? In all of these activities, projection, interpolation and symbolization are the keys to the making and reading of maps. Projecting from one surface to another, making continuous and contiguous what are often discrete and non-contiguous data sets, averaging point data, creating lines and surfaces from sample data, interpreting symbolic imagery in terms of worldly experience: goring, peeling, projecting, selecting, thematizing (Figure 5.1). To me this has always been an act of magic, Merlinesque when done well, like a Monty Python parody when done without skill, craft or commitment.

In the quotations with which I began Chapter 2, Robert Harbison (1977: 124) expressed this issue particularly well:

> From cities of brick to cities in books to cities on maps is a path of increasing conceptualization. A map seems the type of the conceptual object, yet the interesting thing is the grotesquely token foot it keeps in

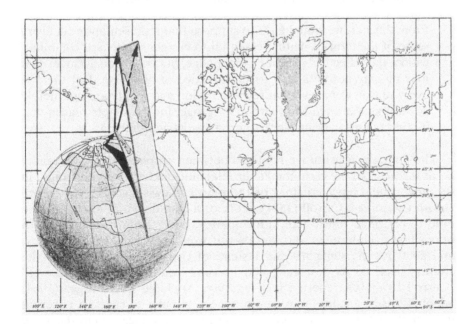

Figure 5.1 Goring, peeling, projecting...

the world of the physical, having the unreality without the far-fetched appropriateness of the edibles of Communion, being a picture to the degree that the sacrament is a meal. For a feeling of thorough transcendence such unobvious relations between the model and the representation seem essential, and the flimsy connection between acres of soil and their image on the map makes reading one an erudite act.

The map is a conjured object that creates categories, boundaries and territories: the spaces of temperature, biota, populations, regions, spaces and objects attain the reality that is particular to them through the combined and multiplied acts of mapping, delimiting, bounding, categorizing – Olsson's drawing and interpreting lines. Maps create objects whose existence is mythic, at least to the extent that these identities are highly formalized abstractions whose effects (once represented as a real object) become very real. Once conjured up, new spatialized identities begin to work as real places and the discourses and practices of cartography and mapping recognize themselves as representing the real. From this vantage point the real has been conjured out of the copper and ink by the cartographic magician: transubstantiation has been achieved and the magician's gold (the real thing) is available to us for use.

But how is this 'real' constituted in the first place and how does it function to allow for the production of discrete identities that have effects? That is, how do geography and cartography produce subjects and identities? In 'Speech and phenomena' Derrida (1991: 9) argues that:

> From the start we would have to suppose that representation (in every sense of the term) is neither essential to nor constitutive of communication, the 'effective' practice of language, but is only an accident that may or may not be added to the practice of discourse. But there is every reason to believe that representation and reality are not merely added together here and there in language, for the simple reason that it is impossible in principle to rigorously distinguish them.

The mutually constitutive relations between representation and reality lie at the heart of the cartographic problematic, and are well illustrated in the now widely repeated story of the map in Lewis Carroll's *Sylvie and Bruno Concluded*. Carroll (1894) wrote of a map drawn at a scale of one mile to one mile which had, unfortunately, never been used because of opposition from farmers who said that 'it would cover the whole country, and shut out the sunlight!'. So, instead of the map, they 'now use the country itself, as its own map, and I assure you it does nearly as well' (Carroll 1894: 169, reported in King 1996: 4). Jorge Luis Borges (1964) adapted this as a story of an empire whose impulse to create a coherent territorial identity for itself led its sovereign to produce a map the same size as the empire. The map was later abandoned to rot in the desert because it was too cumbersome when used (Edney 1997: 1). This Borgean image of a map at the scale of the territory now rotting somewhere in the desert stimulated Baudrillard (1983: 2–3) to problematize explicitly the relationship between map and territory, and to argue that the map precedes the territory, not territory the map:

> It is the map that engenders the territory and if we were to revive the fable today, it would be the territory whose shreds are slowly rotting across the map. It is the real, and not the map, whose vestiges subsist here and there, in the deserts which are no longer those of the Empire, but our own. *The desert of the real itself* ... It is no longer a question of either maps or territories. Something has disappeared: the sovereign difference between them that was the abstraction's charm. For it is the difference which forms the poetry of the map and the charm of the territory, the magic of the concept and the charm of the real. This representational imaginary, which both culminates in and is engulfed by the cartographer's mad project of an ideal coextensivity between the map and the territory, disappears with simulation ...

For Baudrillard (1983: 146) the postmodern experience is one in which the very definition of the real becomes '*that of which it is possible to give an equivalent reproduction* ... not only what can be reproduced, but *that which is always already reproduced*. The hyperreal.'

As we have already seen, Henri Lefebvre (1991: 85) concretizes this notion of the hyperreal when he asks: 'How many maps, in the descriptive and geographical sense, might be needed to deal exhaustively with a given space, to code and decode all its meanings and contents?', and goes on to answer:

> It is doubtful whether a finite number can be given in answer to this sort of question. What we are most likely confronted with here is a sort of instant infinity, a situation reminiscent of a Mondrian painting. It is not only the codes – the map's legend, the conventional signs of map-making and map-reading – that are liable to change, but also the objects represented, the lens through which they are viewed, and the scale used. The idea that a small number of maps or even a single (or singular) map might be sufficient can only apply in a specialized area of study whose own self-affirmation depends on isolation from its context.[1]

At one level, these comments seem rather obvious: mapping the complex spaces of a place or region require many different types of maps each at appropriate scales and each with its own select set of symbols and icons to capture the thematic focus of the coding. But Lefebvre's (1991: 84) point is more complex. It is that comparing different maps of a region

or country and recognizing the remarkable diversity among them illustrates the importance of understanding that '[t]hese spaces are *produced*'.
It is these processes of the production of spaces and the multiple coding of
social spaces that are crucial to understanding the social turn in
contemporary mapping studies:

> Space is never produced in the sense that a kilogram of sugar or a yard
> of cloth is produced. Nor is it an aggregate of the places or locations of
> such products as sugar, wheat or cloth. Does it then come into being
> after the fashion of a superstructure? Again, no. It would be more
> accurate to say that it is at once a precondition and a result of social
> superstructures. The state and each of its constituent institutions calls
> for spaces – but spaces which they can then organize according to their
> specific requirements; so there is no sense in which space can be
> treated solely as an *a priori* condition of these institutions and the
> state which presides over them. Is space a social relationship? Cer-
> tainly – but one which is inherent to property relationships (especially
> the ownership of the earth, of land) and so closely bound up with the
> forces of production (which impose a form on that earth or land); here
> we see the polyvalence of social space, its 'reality' at once formal and
> material. Though a *product* to be used, to be consumed, it is also a
> *means of production*; networks of exchange and flows of raw materials
> and energy fashion space and are determined by it. Thus this means of
> production, produced as such, cannot be separated either from the
> productive forces, including technology and knowledge, or from the
> social division of labour which shapes it, or from the state and super-
> structures of society
>
> Lefebvre (1991: 85)

There are, perhaps, few areas of geography which have historically
come as close to Lefebvre's understanding of the social production of
space and the role played by the state than the emergence of modern
mapping. In what follows, I focus on a reading of mapping practices as
they emerged between the fifteenth and seventeenth centuries, specifically
on cadastral mapping, the state and the economy.

THE EMERGENCE OF A NEW MAP CONSCIOUSNESS

David Buisseret's (1992) *Monarchs, Ministers and Maps: The Emergence of
Cartography as a Tool of Government in Early Modern Europe* and Tom
Conley's (1996) *The Self-Made Map: Cartographic Writing in Early Modern
France* each begin with a simple, but important, question: 'how did it come
about that whereas in 1400 few people in Europe used maps, except for the
Mediterranean navigators with their portolan charts, by 1600 maps were

essential to a wide variety of professions?' (Buisseret 1992: 1); 'Why the sudden birth and growth of mapping?' (Conley 1996: 1).

As we have seen already, to this question Denis Wood (1993) answers that map-making and map use (distinguished from the general ability of mapping and way-finding) emerge with print capitalism and the territorial state. Although the specifics of map development and use are highly complex and regionally differentiated, we can say that in Europe between 1400 and 1600 major changes in the form, use and availability of maps occurred. Moreover these changes – which resulted in a radical trans-formation in map consciousness – were important in influencing (and were in turn influenced by) the emergence of a new national state consciousness whose defining characteristics were a concern for the establishment, defence and management of the national territory, and the administration of the national economy. Buisseret (1992: 4) goes so far as to argue that 'we can be sure that governmental activity was one of the main ways in which Europeans became habituated to the use of maps, and so to the use of new ways of both "seeing" the world and changing it.'

P.D.A. Harvey (1993) has suggested that in England the genre of world maps and regional maps that had become established between 1100 and 1300 had, by the fifteenth century, died out leaving only traces of a medieval mapping impulse in portolan sailing charts and the building plans of stonemasons. Jowett *et al.* (1992: 3) have gone even further arguing that 'with the fall of Rome the use of maps to describe and record landed prop-erty was effectively discontinued' to be replaced by 'written descriptions of the extent of land parcels and their topographical relationships'. In effect, the cartographic imagination of the Roman Empire had been lost, replaced by the chorographical imagination of medieval Europe. Even in Renaissance Italy, where it has often been assumed maps and mapping were commonplace,

> [n]ot a long interest in exploration, nor a long tradition of state ratio-nalization and bureaucratization, nor innovation in the arts and sci-ences, nor a propensity to depict the world 'naturalistically' with linear perspective, nor even the drawn-out military maneuvers during the French invasions of the first third of the sixteenth century left traces in everyday maps in the Italian states.
>
> (Marino 1992: 5)

For Jowett *et al.* (1992: 3–4),

> [s]o complete was the obliteration of map consciousness in feudal Europe that such private property maps as were produced in the medieval world cannot be seen in any sense as survivals of a tradition from antiquity. Property mapping in antiquity is not, therefore, part of a continuous history of the state-sponsored cadastral mapping that

came to characterize European countries in the Enlightenment ... In the emergent capitalist societies of Renaissance Europe, where land became a commodity and power relations were expressed through control of the means of production, which included land, there was now clearly a reason for mapping properties – namely, as an aid to developing the new systems of exclusive rights to land.

Throughout Europe between 1400 and 1600 a revolution occurred in the drawing, distribution and use of maps. Itinerary maps and picture maps (usually from a bird's-eye view) gradually began to disappear and maps of places or areas began to appear in increasing numbers (Figures 5.2, 5.3). P.D.A. Harvey (1993: 8) has suggested that in Tudor England the number of maps remaining from different parts of this period increased rapidly and is a clear indicator of this sea-change. From the second half of the fifteenth century, we have about 12 known maps of particular places or areas; from the first half of the sixteenth century, there are about 200; and from the second half of the sixteenth century there are about 800 such maps. Indeed, 'it is no exaggeration to say that the map as we understand it [of small areas – a house, a field, a town, a tract of countryside, an entire country] was effectively an invention of the sixteenth century' (Harvey 1993: 464).

This invention was enabled by several technical innovations: a standardized scale was introduced into topographical mapping in England in the 1540s; the first printed map in England illustrated the Exodus in a bible produced at Southwark in 1535, and thereafter the production of printed maps grew quickly; triangulation was introduced into England in William Cunningham's *The Cosmographical Glasse* in 1559 adapted from a book published in Louvain in 1533; by the end of the sixteenth century,

Figure 5.2 The earliest woodcut picture of a cartographer at work. From Paul Pfintzing's *Methodus Geometrica*, Nuremberg 1598

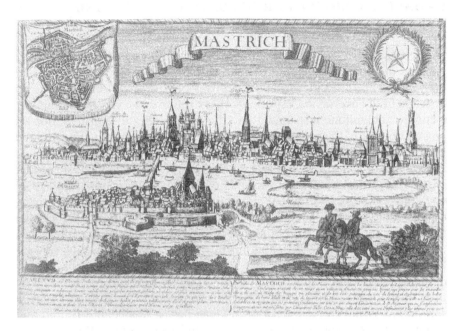

Figure 5.3 Seventeenth-century perspectival view of *Maastricht* by Gravure Sollain
(Plans en Relief: Villes Fortes des Anciens Pays-Bas Français au
XVIIIeS. (1989) Musée des Beaux-Arts Lille, with permission)

the plane-table and theodolite were beginning to be used for mapping; and
in the 1590s legends were beginning to appear on maps to clarify the more
unusual symbols used (Harvey 1993).

Besides these innovations in mapping practice, the birth and growth of
map-making corresponded with (and contributed to) a series of transformations in European Renaissance ways of seeing (especially surrounding the
development of linear perspective, renewed interest in Ptolemaic texts, a
rebirth of interest in quantification and measurement, and new forms of saturated realist painting). But this birth and growth of mapping practices must
also be seen in terms of a series of concrete concerns about property and
identity emerging from political economic transformations of the period.
First, there was a need for maps to envision and consolidate new communities, increasingly imagined as territorially bounded states and discrete unities
of people (articulated in terms of a common history, ethnicity or language
and culture). Second, there was a need for plots and plans for estate planning as private property claims on land and capitalist practices of land
alienation and sale increasingly became the norm. It is these two related elements of the emergence of modern map-making: the role of maps and
mapping in the emergence of new forms of property regime necessitated by
the extension of capitalism, and the emergence of maps and mapping as a
tool of government and state formation, that form the basis for this chapter.

CADASTRES, PROPERTY REGIMES, CAPITALISM AND THE STATE

One primary form of this governmental intervention was the emergence of systems of cadastral mapping; the inventorying and mapping of private land by public authorities for the purposes of governing territory.[2] As an instrument of governance, especially of tax reform, 'the cadastral map was a highly contentious instrument for the extension and consolidation of power, not just of the propertied individual, but of the nation-state and the capitalist system which underlies it' (Jowett *et al.* 1992: 8).

In *Maps in Tudor England* P.D.A. Harvey provides a delightful account of the emergence of new maps in public and private institutions and practice: the military, government administration, urban 'planning', private estates, buildings and the law. But *Maps in Tudor England* is also instructive in another way. It can, I think, be read as an account of the emergence of the discourses, practices and institutions of two competing power–knowledges. The first is that of sovereign power, whose interests in consolidating territorial unity and extending state powers were directly fostered by both the national and the cadastral forms of mapping. At the same time, a second and relatively new form of modern power was being fostered and extended by the new mapping practices and technologies: the power of private property. In so far as cadastral mapping enabled the compilation and dissemination of spatial information on specific places and areas, new economic and political forces emerged able to assert their own interests. Through the expression of those interests they were able to extend the broader processes of economic and political transformations emerging at the time. Lodged between these two, and dependent on each to varying degrees, were the new professions of surveying, mapping, publishing and public administration, for whom the new cartographic practices represented a political and an economic opportunity (in much the same way is currently true of digital information technologies and GIS). As Harvey (1993: 17) argues: 'Cartographic techniques were substantially in advance of the market in Tudor England, ready to be put to use when demand arose. What mattered was the spread of demand, and how map-makers created and fostered this demand for their products.' It was 'this society which, in the course of the sixteenth century, discovered the *value* of maps' (p. 25). This value resided in the ways in which it (like other technologies of the day, especially the printing press) served the needs of several different and often contradictory interests, among them the interests of the crown, the state, the military, the merchants, the private property owners, local communities, publishers and the emerging group of professionals involved in the surveying, mapping and use of cadastral maps.

How precisely was this constellation of interests articulated around cadastral mappings? *The Cadastral Map in the Service of the State* (Jowett *et al.* 1992) provides a series of important answers to this question. Rapidly increasing population generated increases in rival claimants on private and common-resources lands, and new commercial uses created increasing friction among and between existing and new users of common lands. Jowett *et al.* (1992) locate these emerging conflicts firmly in terms of the political economic transformations from feudalism to capitalism, in which cadastral mapping and groups such as professional land surveyors, private estate managers and public administrators were especially important. The new maps and their spokesmen legitimized issues such as precision of location, efficiency of land management and permanence of record as the basis for government interest in mapping.

Beginning in the sixteenth and seventeenth centuries, cadastral mapping became increasingly professionalized and concerned itself more with legal and symbolic issues involved in inventorying private estates, cadastral mapping for tax reform and the provision of general tools of accurate recording for public authorities at the local and state levels of government. In turn, such state-supported cadastres were widely resisted, becoming highly charged and contested political issues.

In the Netherlands, the expansion in the production and use of printed cadastral maps in the sixteenth and seventeenth centuries was directly related to 'its mercantile and imperial expansion during this the Dutch golden age' (p. 44). In the polder areas in particular 'surveying and mapping developed early and rapidly became an indispensable part of public administration' (p. 45). *Waterschap* (polder authority) maps illustrated clearly the ways in which these newly emerging mercantile interests intersected with the interests of the local state to foster new mapping practices. The polder authority maps were needed for the administration, management and accounting of dyke and polder construction and maintenance. From the fifteenth century, the costs involved in these practices had been a public charge and by the sixteenth century the charges had been levied against each village and shared by quota among the villages. Detailed and accurate maps became particularly important for the administration of these levies and, by the sixteenth century, for redressing the unequal burdens that dike taxation by quota imposed. The result was a series of local and regional cadastral maps, some of which were so detailed and accurate as to serve as the basis for tax collection and property records for many years (Figure 5.4). Schillincx's 1617 map of Putten, for example, served as the standard map for tax collection in the town for over 250 years (Jowett *et al.* 1992: 14). Besides serving the administrative and fiscal needs of polder management, the cadastral maps also had an important public relations and advertising function. Especially in the seventeenth century, when a surplus of capital and land drew investments from urban merchants and manufacturers who wanted to diversify their investments, cadastral maps served as publicity tools for reclamation schemes. These 'merchants' drainage projects were thoroughly capitalist undertakings' (1992: 20) in which maps were deployed from conception, in the planning

Figure 5.4 Platted spaces of the perfect state. *Lille avant les traveaux de vauban.* Gravure de Blaeu, 1649 (Plans en Relief: Villes Fortes des Anciens Pays-Bas Français au XVIIIeS. (1989) Musée des Beaux-Arts Lille, with permission)

and construction phases, to the allotment of plots to the decorative representations for investors of the plots they had purchased.

In Sweden, Finland and Norway the situation was different: subsistence production in peasant households was much more important than commercial agriculture (p. 47). Here the imperative of the central state under the power of the monarchy was to consolidate the power of the central state in part by developing a national taxation system. To this end, the development of national mapping programmes was encouraged from the sixteenth and seventeenth centuries onwards. In Denmark, the establishment of an absolute monarchy in 1660 and its attempts to consolidate its economic and political power and diminish the economic and political power of the nobility – in part through taxation policy – also gave an important stimulus to a comprehensive surveying and mapping programme (p. 116). In all the Nordic countries, open-field enclosures and the transfer of common property rights to individual owners gave added value to the project of detailed cadastral mapping, creating its own base of support even in the face of the centralization of state power (p. 117). Jowett *et al.* (1992: 118) are very clear about this: 'Where private property rights are not established, the development of cadastral mapping is problematic if not impossible.'

In part, this difficulty or impossibility stems (for Jowett *et al.*) from the resistance of those who own the land collectively in terms of common rights. But in part it stems from very different notions of value under such property regimes. For example:

To have mapped ownership of *skyld* [a measure of the value of land rather than the land itself, and to which – not to land – individuals could lay claim] would have been impossible, just as today mapping ownership of shares in a company is impossible.

(p. 118)

Only with the transfer of individual land plots to individuals and the assignment of ownership rights did cadastral mapping become possible. But this was not a one-way relationship, as the use of maps also resulted in significant changes in the concept of ownership and the meaning of land. With maps came the dissemination of standardized measures and a simplification in the types of measure: thus measures of land quality (like *skyld*) were substituted for by land area, and land price then became a surrogate for concepts of land value and quality. In Livonia, for example, the *uncus* was a cadastral not a real measurement of land, and took into account quality, proximity to market, associated labour requirements and customary dues. But, with the imposition of the Swedish land surveys, a simple and uniform measure of land area was substituted (Jowett *et al.* 1992: 118–19).

In Germany the situation was much more complex, in part because of the legacy of war and in part because of the fragmented nature of the *Länder*. Here, war, the need for symbols of territorial identity, and the demands for 'administrative reforms and mercantilist state direction of the economy' were all important factors in the development of cadastral mapping, and '[w]ith the development of the territorial state, rulers wanted to get as full a picture as possible of the extent and condition of their territories and found the map a useful instrument of this task' (Jowett *et al.* 1992: 168).

Austria and Italy provide fascinating counterpoints to this story of national mapping projects on the part of a hegemonic state. Cadastral mapping was instituted in Austro-Hungary as a whole only in the nineteenth century. Until that time, any effort to consolidate the power of the Habsburg state at the expense of the nobility and church was fiercely resisted. In Italy, cadastral mapping was resisted unsuccessfully in Milan and successfully in Tuscany, and opposition ensured it was not attempted in the remaining territories (Jowett *et al.* 1992: 203–4). Cadastral mapping in France, by contrast, was particularly important for reforming tax policies – a necessity for Louis XVI and a central objective of the French Revolution.[3] Indeed, in 1929 Marc Bloch suggested: 'Tax reform was one of the *raison d'être* of the revolution: to base taxation on land in a manner as equitable as possible, topographic surveys were absolutely essential.'

The *plan terrier* compiled for Corsica in the 1770s and 1780s distinguished Crown, common, and individual land parcels, and had important implications for economic development beyond fiscal reform. This fact was recognized by the post-revolutionary government under Napoleon, who leant strong support to the development of the *cadastre parcellaire* (Jowett *et al.* 1992: 205). In fact Napoleon had long been a supporter of detailed mapping programmes and had used large-scale miniature plans and models in military campaigns for many years to great effect (Musée des Beaux-Arts Lille 1989) (Figure 5.5).

There is some disagreement about the dates when a broad-scale mapping consciousness emerged in Britain. Harvey (1993: 7) argues that 'in the England of 1500 maps were little understood or used. By 1600 they were familiar objects of everyday life.' But Jowett *et al.* (1992: 263) suggest that a national cadastral mapping programme did not emerge until much later. Crown and parliamentary land surveys from the seventeenth century contain few or no mapping provisions, and throughout the sixteenth and seventeenth centuries land enclosures had occurred in many parishes without recourse to maps. Only with the Enclosure Acts of the late eighteenth century were maps required (Jowett *et al.* 1992: 263). For our purposes, however, the point is clear: between the sixteenth century and the eighteenth century, maps became a fundamental part of everyday life and the practices of the state.

Perhaps the most thorough investigation to date of the role of mapping in the practice of statecraft is Tom Conley's (1996) *The Self-Made Map: Cartographic Writing in Early Modern France.* Conley (1996: 2) begins his book with the wonderful wordplay on 'self-making' in the title – both the production of the map *and* the making of identity (the national, French, citizen self): 'a theatricalization of the self, which acquired a consciousness of its autonomy through modes of positioning that are developed into both textual and gridded representations of reality'. Conley points to the ways in which the sudden birth and growth of mapping – 'a new cartographic impulse' – emerges along with early modern print culture and an emerging sense of national identity.

He begins with the same question posed by David Buisseret: why, between the early fifteenth and seventeenth centuries, was there such a 'sudden birth and growth of mapping?'. He gives several reasons by way of an answer. The Renaissance rediscovery of Ptolemy triggered a surge in innovations in printing, especially in wood-cutting, copper engraving and movable type, and these not only facilitated the emergence of a modern print culture (and as Anderson reminds us, a print capitalism), but also mapping culture and mapped capitalism. Developments in science and technology placed added emphasis and value on quantification and measurement. Innovations in the visual arts, especially the emergence of 'saturated realism' in Northern Europe and one and two point perspective in Southern Europe, prioritized the 'naked eye' over other senses as the basis for empirical 'observation'. These were further enhanced by the use of mapping in estate planning as private property regimes were extended and public resources were redefined and enclosed. The projects of political unification, nation building and the consolidation of a notion of national space gave added importance to mapping projects, particularly in regard

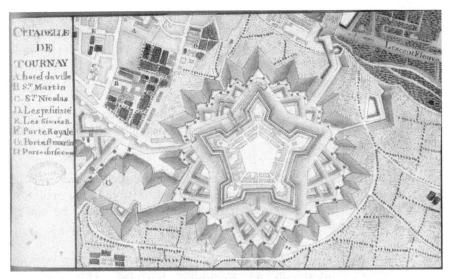

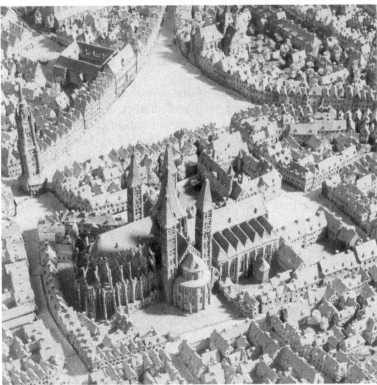

Figure 5.5 Citadelle de Tournay [top] and photograph of *Le Plan en Relief de 1701, Tournay* [bottom] (Plans en Relief: Villes Fortes des Anciens Pays-Bas Français au XVIIIeS. (1989) Musée des Beaux-Arts Lille, with permission)

to the defence of territorial borders. The national origins of early modern print culture are thus paralleled by (and related to) the emergence of mapping culture. A new cartographic impulse thus emerged historically along with a newly emerging sense of national identity, and – as Denis Wood insisted – from that point onwards, map-making is effectively a form of statecraft.

The spatial strategies of nationhood and selfhood emerged in books on navigation, island atlases, sheet maps, cosmographies and atlases, and Conley (1996: 6) claims:

> 'Selfhood' and 'self-fashioning', and their consequent import on the creation of national subjects become especially visible in the evolution of cartographic writing from the years of humanism to the age of Henry IV and the subsequent growth of French cartography.

The cartographic impulse is, then, a means of coping with the growing recognition of finitude in an expanding universe. This emerging autonomous selfhood is: 'A drive to locate and implant oneself in a named space; a drive to imagine necessary connections between the "I", the locale of its utterance, and the origins of its birth ... a perceived need to burrow into and circulate about a body, a world, and a nation' – giving credence to an illusion of origins (p. 303). Conley points not only to the intense interweaving of mapping with national and self-identity, but to the fact that the issue of gender, origin, *eros* and identity are produced cartographically through the mapping of the spaces of nation and nationhood.

But, if cartographic writing infused the making of national identity in Europe, it was deployed universally in the non-European world to decode existing social and territorial structures and to forge a modern national body, what Thongchai called the geo-body of the nation. It is to this notion of the nation building and national identity in the non-European world that I now turn.

6 Mapping the geo-body
State, territory and nation

In terms of most communication theories and common sense, a map is a scientific abstraction of reality. A map merely represents something which already exists objectively 'there'. In the history I have described, this relationship was reversed. A map anticipated spatial reality, not vice versa. In other words, a map was a model for, rather than a model of, what it purported to represent ... It had become a real instrument to concretize projections on the earth's surface. A map was now necessary for the new administrative mechanisms and for the troops to back up their claims ... The discourse of mapping was the paradigm which both administrative and military operations worked within and served.

(Thongchai, *Siam Mapped: A History of the Geo-Body of a Nation*)

POWER TALK AND THE PRODUCTION OF NATURE

The drive for overseas exploration, knowledge, and wealth (always through some form of national competition) accelerated the technologies of modern mapping and the territorialization of non-European lands. In the preceding chapter, we have seen how European states became involved in and committed to the development of national mapping programmes, particularly to consolidate the power of the central state against sectional and regional interests. As in the case of cadastral mapping, the response on the part of the people of the regions was mixed; new private owners and investors supported these means of defining accurately and legally the boundaries of their private property; large segments of the population supported tax reform which addressed regional and social inequalities; and established feudal interests resisted, in some cases strongly and successfully, the extension of the power of the central state and by implication the diminution of their own powers (as in the provinces of Austro-Hungary). However, the geo-coding of the body politic went hand in hand with the extension of state interests and private property regimes, with the result that local knowledges and valuations, regional systems of topophilia, and alternative mapping opportunities were eradicated or sublimated under the universal logic of law, administration and measurement.

Perhaps the boldest of these territorializations was that created by the Papacy in 1494 when, in delimiting the expansionist zeal for overseas wealth, the Tordesillas line was drawn across the Americas to both delimit and open up the overseas empires of Spain and Portugal (Figure 6.1). By the late nineteenth century, this paradigmatic discourse of mapping had so informed strategic thinking about state and territory that the geopolitical practice of empire took on what, in hindsight, seems like an increasingly arrogant cartographic imagination. As Tom Bassett (1994: 316) has shown, this imagination was at work at the Berlin Conference of 1884–5, which convened the leaders of the major European states to negotiate the African territories before war broke out over them:

> huddled around a map of the continent as they drew boundaries of their purported possessions and spheres of influence. That carto-graphic partition of Africa inextricably linked map-making and empire building. Yet the act of drawing lines on a map is only one example of how cartography furthered imperialism. Maps were used in various ways to extend European hegemony over foreign and often unknown territory.

Bassett goes on to elaborate the ways in which nineteenth-century mapping practices formed an integral part of the political discourse that fostered and supported the colonization of Africa. Maps were used to promote and assist European expansionism and then, once colonization began, cartographic techniques were used to further the imperial project. This project was particularly clearly illustrated by the figure of Cecil Rhodes astride the African continent (Figure 6.2), but it was nowhere clearer than in the land surveys in the United States of America.

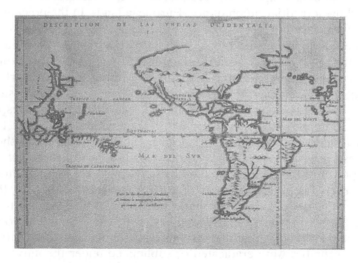

Figure 6.1 Treaty of Tordesillas, 1494

Figure 6.2 Cecil Rhodes – From Cape to Cairo

In 1976 Hildegard Binder Johnson published *Order upon the Land*, an account of the development, imposition and effects of the rectangular land survey in the Middle West of the United States. Strikingly similar to Lefebvre in its own claims about the production of space, *Order upon the Land* graphically illustrates the importance of cadastral mapping in framing and producing the cultural landscapes and geographies of private property in the United States (Figure 6.3). The map has been an integral tool in the structuring and functioning of what we take to be the everyday and the natural: the agricultural landscape, the modern city, the road along which we drive, the very rooms in which we sit and read, and even the processes by which the worlds of public safety, basic infrastructure, and commerce enter our towns, neighbourhoods and houses are all products of the working of maps and the institutions and practices that support their construction and deployment.

In consort with other technologies of social control, such as violence, disease and alcohol, mapping programmes contributed to the physical eradication and historical erasure of indigenous populations. In this light, Robert Rundstrom (1991: 1) has argued for a widening of the scope of a postmodern cartography to avoid what he calls the 'unacceptable restrictions on what maps and acts of mapping we study'. Even Harley's deconstructive reading focuses too narrowly on the map as text and fails, as a result, to pay attention to acts of indigenous mapping in which process, for example, might be more significant than textuality. In non-textual mapping cultures, there 'has been a quite different crisis of cartographic representation' (Rundstrom 1991: 2), marked by the need to resist the representational, textual economies of the West (M–C–M – military–capital–map). The study of such knowledge–constitutive interests and the

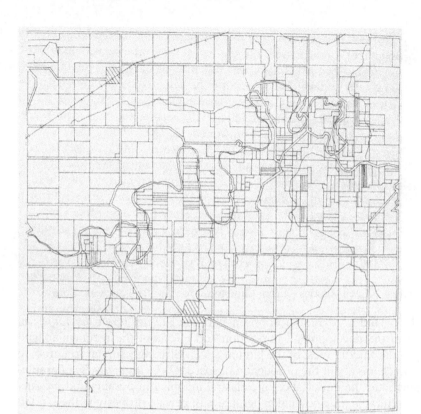

Figure 6.3 Gridded lands: lines, landholdings, landscapes. From Hildegard Binder
      Johnson, *Order upon the Land* (with permission, Oxford University Press)

political and cultural economy of text-based societies requires a study in
ideology critique. In one sense, this is what *Order upon the Land* provides.
As Johnson (1976) puts it so gently:

> Sixty-nine per cent of the land in forty-eight states is contiguously
> covered by the rectangular survey, and 9 per cent is intermittently
> covered in the remaining area, including Alaska. Of the 1.8 billion
> acres at one time in the public domain, approximately 1.3 billion have
> been surveyed. In the vast literature about the frontier and agricul-
> tural history of the Upper Middle West the survey is taken for granted
> and is generally accepted as an advantage for settlement. For example,
> Frederick Jackson Turner, after extolling the Ordinance of 1787 in an
> article on the Middle West (*International Monthly*, December 1901),
> stated: 'The Ordinance of 1785 is also worthy of attention ... for
> under its provision almost all of the Middle West has been divided by
> the government surveyor into rectangles of sections and townships by

whose lines the settler has been able easily and certainly to locate his farm and the forester his forty'. *In the local organization of the Middle West these lines have played an important part.*

(Johnson 1976: iii; emphasis added)

Scott Kirsch (1999) has recently suggested that regional surveys and mapping were successful precisely because of the ways in which violence was erased and excluded. In the government surveys of the American West, '[t]he spatial ordering of information that could be shown on a map was not merely a reflection of what was already "there" to be surveyed in the West' (p. 3). Surveyors like John Wesley Powell 'brought with them a particular way of seeing the land and its inhabitants' (p. 3), and his field surveyors were explicitly charged by Powell to triangulate their surveys with established frameworks and pre-existing surveys; to consult existing Public Lands Surveys in order to connect 'the established line with your system of triangulation' in order to reproduce existing mineral claims. Indian lands, agricultural lands, and mineral lands, in particular, were to be mapped in the interest of state, nation and territorial development. Thus, Secretary Delano of the Department of Interior writing to Powell on 1 July 1874 insisted:

It *will* be borne in mind that the ultimate design to be accomplished by these surveys, the preparation of suitable maps of the county surveyed, for the use of the Government and the nation, which will afford full information concerning the agricultural and mineral resources and other important characteristics of the unexplored regions of our Territorial domain.

(quoted in Kirsch 1999: 4; emphasis added)

In this sense the map is a hidden (or not so hidden) tool – a plan – for delimiting the environment and the practices that take place in it. But it is also an explicit tool for the transformation of social, economic and political spaces of the state. In the case of Powell, the western land surveys offered an alternative to the use of military force. The 'Indians' west of Colorado could, he thought, be subdued at less cost and without the application of military force through the deployment of 'a system of social control through spatial strategies of concentration and heightened visibility'. At the heart of this system of social control and visibility was the survey. The map and the institutions within which it was produced, functioned as an archetype of what Michel Foucault referred to as a power–knowledge. That is, a discourse, practice and set of institutions that delimit potentialities through the control of space–time–action and thereby produce certain types of subjects, actors and places. Power–knowledge is thus a form of power that is at the same time both delimiting–controlling *and* enabling: that is, it is a form of productive power.

We now have many detailed accounts from the eighteenth and nineteenth centuries of the geopolitical role of maps and the history of spaces of which Foucault wrote (see, for example, Hannah 2000 and Dorn 2002). But surprisingly few such accounts have been written for the twentieth century (see also Monmonier 2001). Perhaps there exists few starker illustrations of the role mapping has played in the constitution of geographical and social identity in the twentieth century than that which took place behind the scenes at Yalta. In his autobiography, Charles Bohlen (1973: 152) described the ways in which the map and geographical imagination functioned at the Teheran Conference of December 1943 between Stalin, Churchill and Roosevelt.

> At the final plenary session that evening, Roosevelt did not take part in the discussion regarding Poland. As Churchill and Stalin talked over the problem, I noticed that the British and the Russians were working on a map of Poland torn from *The Times* of London. Since we had brought a collection of books with various maps touching on the Polish issue, I asked the president whether he would have any objection to my lending a copy to Stalin and Churchill to make their discussion easier. The president gave me permission, and I took a book over to Stalin, who looked at one map and asked me on what data these lines had been drawn. The map showed the ethnic divisions of eastern Poland. I informed the marshal that as far as I knew, the only data available came from Polish sources. Stalin grunted and took his ever-present red pencil and somewhat contemptuously marked the map to show what would be returned to the Poles and what would be kept for the Soviet Union.

This passage is stunning in its depiction of the cavalier and, as Bohlen writes, contemptuous manner in which fundamental decisions about the future of Europe were made (the use of a map torn out of the newspaper as the basis for delimiting boundaries, the scribbling of lines on large-scale maps to demarcate national territories, and the ceding of responsibility for Poland to the Soviets without comment) and on the basis of which the next fifty years (and longer) would be shaped. The free play of geopolitical imagination becomes even clearer in the next paragraph from Bohlen (1973: 152):

> During their discussion, Stalin and Churchill virtually agreed on the future borders of Poland. The frontiers included the Curzon line in the east, with modifications as Stalin had indicated, and the Oder–Neisse line in the west. In other words, the new Poland would give up Poland's eastern areas to Russia in return for parts of eastern Germany. This understanding, which was entirely oral, led to further confusion later on because there were two rivers Neisse, a western and

an eastern, and there was no mention of which one they were talking about. The division that Churchill and Stalin agreed to is the one that still exists.

It is precisely through the map's objectivity that particular representations of society and nature are materialized; the interests of one or another group are inscribed while other interests are erased or silenced. But, in part precisely because of such clear examples of gross cartographic hubris such as those demonstrated at the Berlin and Teheran Conferences, this deconstructive turn has itself 'turned' quite rapidly into a reading of power relations into social relations and maps. As David Matless (1999: 193) sees it:

> Critical analyses of cartography have aligned maps with an impulse to dominate: land, people, things, properties, colonies. The cartographic eye is equated with the eye of power-as-domination. Alternatively maps are claimed as a vehicle of resistance, a language whereby rights to place may be asserted or through which non-dominatory representations might be cultivated.

Yet, as Matless realizes, such a duality of approaches seems unsettling; much too grounded in voluntarism and liberal individualism. And so, for the moment, I want to interrogate this duality briefly before returning to Foucault, and I want to do this by considering Brian Harley's work again.

The admixture of deconstruction and ideology critique in Harley's work has unleashed what I shall call here 'power talk'. 'Power talk' in cartography and geography has become ubiquitous. At one level, as I have indicated, this is a very positive turn for a field that only fifteen years ago could not understand why questions of power, inequality and justice might be part of the 'scientist's' bailyiwick. On the other hand, this turn to issues of power is itself a kind of power 'talk' and we need to be wary about the kinds of functionalism that reduce the map and mapping enterprise to a mere instrument of the powerful and devious. We need also to heed the warning given by Michel Foucault in *The History of Sexuality Volume 1* about interpreting modern power in terms of any kind of repressive hypothesis.

Such a reductionism – 'power talk' – is common, often turning on a particular phrasing. Tom Bassett's (1994: 316) otherwise important reading of the geopolitical cartographic imagination in Africa with which I began this chapter falls subject to this reduction to 'power talk' when he renders the map as a direct and unambiguous instrument of power:

> From a theoretical perspective, this study views cartographic truth as an example of an exercise of power, linked to the will to dominate and control (Foucault 1980, 131). Because maps are an expression of the 'territorial imperatives of a particular political system', they are both instruments and representations of power (Harley 1998a).

There are, of course, important reasons for stressing the ways in which symbolic representations operate as instruments of power, and how they mask the underlying ideology that produced them or the social interests that supports them. We have seen this above and this was precisely what I tried to do in my work on GIS and the surveillant society, and what Neil Smith tried to do when he argued that the Gulf War was the first GIS war. The stakes were/are high and – particularly as the first high-level pin-point bombing raids using smart bombs were unleashed in 1991 on thousands of targets across Iraq – there was a political and scholarly responsibility to speak clearly and directly about the unsettling fact that mapping technologies and practices were important facilitators of such weaponry. As mapping unleashed possibilities for new rounds of military horror and violence it was important to challenge those arenas that refused to acknowledge or reflect upon their own complicities and commitments.[1] In a similar way, Bassett and others have refocused our attention away from fetishized notions of maps as antiquarian or technical objects, centring attention squarely on the living consequences of map use.

Responses to such work on the social implications of mapping and GIS have often focused too one-sidedly on repressive notions of power, assuming that any discussion of the power of maps or GIS must be criticizing their uses and/or their users for being tools of power. While it is important to ask who sets the agenda, who provides funding, and what are the actual broader institutional arrangements within which modern cartography, mapping practices, GIS and science generally function, this is only one element in the project of social critique in which we locate mapping as a form of power–knowledge. Mapping as a power–knowledge also functions as productive power to constitute objects, identities and practices that are part of (and constitutive of) our world. In this sense, we want to know how maps do work. Instead of asking the more traditional (and problematic questions) 'what do maps represent and how do they do so?' we ask instead 'what effects do such representational practices have in the world, and what work is achieved by claims to representational accuracy?' Just as we might want to know how the Federal Reserve or smart-transport systems or telecommunications work as related sets of technologies, discourses and institutions that 'act' and do work, we need to understand how the discourses, institutions and practices of mapping do work in the world. But in so doing we need an understanding of mapping that does not reduce the work maps do to the repressive exercise of power. In moving beyond 'power talk', we need some way of understanding the constitutive role maps play in shaping identity and practice. The rest of this chapter focuses on the ways in which mapping practices have shaped conceptions of nature, how a particular planetary consciousness emerged from a multitude of alternative possibilities, and how this scientific notion of nature has of late begun to break apart. In particular I seek to destabilize the 'power talk' of domination and resistance by looking for examples of where the erasure of indigenous peoples from the land and the map (their crisis of representation) can *also* be thought in terms of the heteroglossic spaces of transcultural mappings.

TRANSCULTURAL MAPPING AND THE GEO-BODY OF THE NATION

In deconstructing the 'power talk' of the power of maps, we need to ask about the ways in which hegemonic visions were *in fact* constructed and disseminated, and to question the metaphors of imposition, overlay and eradication. Actually existing modern hegemonies are, or were, rarely hegemonic in an absolute sense, and this was precisely why Foucault (1979) distinguished between sovereign power and modern power in *Discipline and Punish*. Modern hegemonies are complex interweavings of coercion and consent, as Gramsci (1981) suggested, and in these inter-weavings bodies, subjects and identities are produced and inscribed in multiple ways and forms.

Deleuze and Guattari (1983, 1987) have described at length and in great detail the many ways in which such decodings and recodings shape the particular form of the *socius*, but also how such codings also always fail to capture the lines of flight of what they call 'desiring machines'. In the production of productions, inscriptions and consumptions, maps have played a fundamental role of decoding, recoding and over-coding nature, space and the *socius*. This is, I think, the same impulse we find in *Mapping Men and Empire*. Here Phillips (1997: 5) documents the ways in which imperialism went hand in hand with mapping as an enterprise of naming and possession, charting the world and then colonizing it. But he also stresses that 'while geographical imaginations and narrative adventures often appear committed to "continuous reinscription" of dominant ideologies of masculinity and empire, the geography of adventure is neither deterministic nor static.' Indeed, he goes on, alongside such reinscriptions of dominant ideology are 'points of departure' leading in entirely different and transgressive directions. Against functionalist and reductionist readings of adventure writing and mapping, we need also readings open to these transgressive moments. It is to the first of these that we now turn.

Triangulations, land-surveys and military defence

Benedict Anderson (1991: 170–1) has suggested that:

> Cairo and Mecca were beginning to be visualized in a strange new way, no longer simply as sites in a sacred Muslim geography, but also as dots on paper sheets which included dots for Paris, Moscow, Manila and Caracas; and the plane relationship between these indifferently profane and sacred dots was determined by nothing beyond the

mathematically calculated flight of the crow. The Mercatorian map, brought in by the European colonizers, was beginning, via print, to shape the imagination of Southeast Asians.

Colonial and later national military surveyors in Southeast Asia were 'on the march to put space under the same surveillance which the census-makers were trying to impose on persons. Triangulation by triangulation, war by war, treaty by treaty, the alignment of map and power proceeded' (Anderson 1991: 173). For Thongchai (1994), it was precisely such techniques and practices of mapping that constituted the nation of Siam. Mapped, materialized and given form, Siam could function as a juridical–political entity – a nation-state.

> In terms of most communication theories and common sense, a map is a scientific abstraction of reality. A map merely represents something which already exists objectively 'there'. In the history I have described, this relationship was reversed. A map anticipated spatial reality not the vice versa. In other words, a map was a model for, rather than a model of, what it purported to represent ... It had become a real instrument to concretise projections on the earth's surface. A map was now necessary for the new administrative mechanisms and for the troops to back up their claims ... The discourse of mapping was the paradigm which both administrative and military operations worked within and served.
>
> Thongchai (1994: 310)

From the fifteenth century onwards, political struggles to consolidate control over fiscal and other policies in the hands of the central state, wresting power from sectional and regional interests was also crucial to the construction and constitution of the apparatus of the newly emerging territorialized national states and nascent national (capitalist) economies of the South. Here, cadastral mapping was a particularly important tool in this process of coding territory as private, calculable (hence taxable and tradeable), and part of a larger territorialized entity that was the rightful domain of government. In the process, cadastres created standardized systems of symbol and measure – a statist and capitalist coding – which subsumed local and regional differences in land practice, erased topophilic forms of value (*skyld*, for example), and in their place in time established a universal language, not only of national forms of speech, but also of land and territory.

Thongchai Winichakul (1994) has called this process of nation building on the periphery of European Enlightenment and modernity the construction of the geo-body of the nation; the mapping of the nation-state on to (and over) the cultures of difference and identity that preceded the geopolitics of the national imagination. In *Siam Mapped*, he shows how the imperial eye gridded and categorized the non-European subject in ways that brought a kind of visual and cognitive order to the European encounter with the non-West, an order that was also enforced and policed with truncheon and Gatling gun. But Thongchai also insists that the discursive construction of what it was to be Siamese and the notions of nationhood that attended its emergence should not be seen as a disembodied or voluntarist act of will or consciousness. For Thongchai (1994: 15), Benedict Anderson's *Imagined Communities* 'seems too concerned with the imagination, the conceivability of the nation'. Thongchai added to it an analysis of the ways in which maps created the geo-body of the modern nation. Maps operated as technologies of territoriality and were constitutive of Siamese nationhood: 'nationhood was', he argues, 'figuratively created by our conception of Siam-on-the-map, emerging from maps and existing nowhere apart from the map' (Thongchai 1994: 17).

This mapped identity was very different from that of indigenous Southeast Asian tradition. Here 'a subject was bound first and foremost to his lord rather than to a state. People who lived in one area might not necessarily belong to the ruler of that area, although they might still have to pay tax or rent to the lord on that land ... it was a peculiar custom in which the power over individuals and land was separated' (Thongchai 1994: 164). The differential system of filial commitments and rights over land – as with the complex evaluations of land quality in Dutch traditions of *skyld* and *ensuur* – posed serious problems for the modern administrators who wanted to fix national identity in terms of territoriality. The 'nomad' has long served as the 'pariah' form of the naturalized subject. For the nomad, as well as for many agrarian and trading peoples, the notion of belonging to specific territory, bounded by a fixed political boundary defines and delimits the possibilities of identity in ways that are anathema to them. For the naturalized national subject, identity is given through the territorialized structures of the state, nation, citizenry, economy and apparatuses of administration, a fact long recognized by colonial administrations around the world, and nowhere better illustrated than in Matthew Edney's (1997) book on the Great Trigonometric Survey and the geographical construction of India.

As Matthew Edney (1999: 1) has pointed out: 'Imperialism and mapmaking intersect in the most basic manner'. Knowledge and territory are inextricably linked in the imperial and state projects, for to establish, police, and control boundaries one must first have 'a certain perception of geographical space' (Nicolet quoted in Edney 1999: 1). The British in South Asia knew this and, through the military and civilian officials of the East India Company, they 'undertook a massive intellectual campaign' (p. 2) to map the territory and thereby define the spatial image of the Company's empire. As Edney (1999: 2) asserts: 'The empire exists because it can be mapped; the meaning of empire is inscribed into each map.'

In this sense, the Great Trigonometrical Survey of India was much more than a project of territorial mapping. The spaces created by the survey were coherent, geometrical, accurate, and uniformly precise; a rational space for the ordering of an imperial archive, holding 'the promise of a perfect geographical panopticon' (p. 319). This rationalizing of space and time structured a similar kind of 'order' and with it a determinate 'visibility' of the social throughout the colonial world. From India to Egypt the 'exotic' was made visible for Europeans by all manner of representational technologies. The map was one of the principal of these (Edney 1999, Mitchell 1991). The Great Survey literally and figuratively inscribed India as a territorially bounded object, reducing it to a rigidly coherent, geometrically accurate and uniformly precise imperial space. Through training in the lower-level surveying technologies, Indians were to receive some of the intellectual skills needed to preserve European rationality (Edney 1997: 319). Reduced in this way, India in all its geographical aspects was rendered knowable to the British.

In the process, the Great Trigonometrical Survey also recoded Britain and British national identity. As the eighteenth-century grand surveys consolidated and disseminated a new lexicon, the trigonometrical surveys introduced new notions of accuracy and space into the domestic political economy. The claim that small-scale maps were based on 'actual surveys' had already emerged as a discursive device for 'correctness and verisimilitude'. By the 1820s, any such claims required trigonometrical surveys 'to be proper and correct' (p. 320). The authority of the sciences of mapping – survey, accuracy, verisimilitude – was bolstered by the imperial project of colonizing and subjugating India. In turn, the authority and discourses of rational method, spatial ordering and social panoptical control returned home to Britain – what Peter Gould used to refer to as the waves of recolonization of the British by their own systems of colonial administration. In this sense, there is a delicious ambiguity to Edney's (p. 340) final sentence: 'The empire might have defined the map's extent, but mapping defined the empire's nature.' It did indeed, and it did so at the heart of the empire as in its dominions, as we will now see.

Mapping as heteroglossia and transculturation

In *Imperial Eyes: Travel Writing and Transculturation*, Mary-Louise Pratt (1992) interrogates the geographical and cartographical imagination of Alexander von Humboldt through his thirty volumes of travel and nature writings on the Americas. She illustrates at least three issues that are important for our own consideration of the cartographic impulse. First, through an analysis of the reception of von Humboldt's writings (enthusiastically so by Charles Darwin and Simon Bolivar, for example), Pratt demonstrates the ways in which geographical writing and mapping were crucial in framing a particular identity for the Americas, both in terms of Europe's experience of the 'new world' and in terms of Spanish America's own self-understanding. Second, she shows how European geographers such as von Humboldt mapped the contact zones of the 'new world' by erasing local peoples and their histories and inscribing maps and geographies of primal nature in their place. These mappings were on the one hand constitutive of the geographical imaginary that founded European colonial and imperial designs, and on the other hand went largely unacknowledged as a form of political/imperial practice that was functional to Euro-expansion. And third, Pratt shows how, despite these erasures, European geographies and mappings of the contact zone were what she calls 'transculturated images', derived from and saturated with local knowledges and imagery, and reflecting the heteroglossic not monolithic structure of colonial space.

Von Humboldt's encyclopedic impulse (between 1805 and 1834 his writings, drawings, and maps filled thirty volumes) aimed at completely filling the northern European map of Spanish America, and as more and more Europeans began travelling to South America, von Humboldt 'remained the single most influential interlocutor in the process of re-imagining and redefinition' of post-colonial Spanish America (Pratt 1992: 111). His reinvention of South America 'first and foremost as nature' (Pratt 1992: 120) was an explicit attempt to go beyond (and offer a challenge to) the taxonomic sciences of nature of the Linneans, and in its place to project 'a dramatic, extraordinary nature, a spectacle capable of overwhelming human knowledge and understanding ... a nature in motion, powered by life forces, many of which are invisible to the human eye; a nature that dwarfs humans, commands their being, arouses their passions, defines their powers of perception' (Pratt 1992: 120). This sublime rendering of a nature that overwhelmed (and erased) the traces of human beings in the landscape, introduced European readers to a new type of nature discourse. In the process, his abstractions largely erased local culture and life from the images of South America: Humboldtian Euro-expansionist imperialism was to be achieved through a reimagining and reordering of Spanish America as raw nature, available resource and empty territory. As he argued in 1814,

> In the Old World, nations and the distinctions of their civilization form the principal points in the picture; in the New World, man and his productions almost disappear amidst the stupendous display of wild and gigantic nature. The human race in the New World presents only a few remnants of indigenous hordes, slightly advanced in civilization; or it exhibits the uniformity of manners and institutions transplanted by European colonists to foreign shores.
>
> (Von Humboldt 1814, quoted in Pratt 1992: 111)

While the discourse of nature was re-mapping South America, it also sought to reframe bourgeois subjectivity by providing an alternative to its strategies of separating objectivism and subjectivism, knowledge and experience, science and sentiment in what Pratt (1992: 119–20) calls 'a new kind of planetary consciousness' (Figure 6.4): 'Humboldt's brand of planetary consciousness makes claims for science and for "Man" considerably more grandiose than those of the plant classifiers who preceded him. Compared with the humble, discipular herborizer, Humboldt assumes a godlike, omniscient stance over both the planet and his reader' (Pratt 1992: 124).

In this sublime grand perspective, the human dweller was erased and a primal nature was 'codified in the European imaginary as the new ideology of the "new continent"'. In practice, von Humboldt and his travelling companions belied the very goals of their planetary consciousness. They relied totally on the networks of 'villages, missions, outposts, haciendas, roadways, and colonial labor systems to sustain themselves and their project, for food, shelter, and the labor pool to guide them and transport their immense equipage', and for a traveller who 'never once stepped beyond the boundaries of the Spanish colonial infrastructure', von Humboldt's coding of Spanish America as primal nature can only be seen as ironic. In his writings, sketches and maps these systems and networks of economy and community were largely ignored or, where present, were rendered as in the service of Europeans or as potential resources for the civilizing influences of European capital and industry. The effect was a deterritorializing of indigenous peoples and a denial of their history and culture: a remapping of lived space

Figure 6.4 Mount Chimborazo Flora, Topography and Altitude Mapping, Alexander von Humboldt (Alexander von Humboldt, *Geography of Plants*, 1807)

as primal nature, and the invention of a geopolitical mapping of nature and society. And it was achieved through a project of analytical abstraction.

Von Humboldtian cartography involved the making of what Anne Godlewska (1999a: 236) calls a 'new kind of map'. A transitional figure from classical to modern *episteme*, von Humboldt reflected the eighteenth-century's preoccupation with accuracy, numeracy and measurement, but rejected the naïve empiricism of the time with its 'overwhelming preoccupation with measuring and mapping "anything and everything" ' (p. 244). Von Humboldt insisted that he was 'attempting an abstraction' (p. 245), not in pursuit of enhanced accuracy, possession and representation of 'nature' – but in order to create a nomadic abstraction that would be transferable across fields and space. Godlewska (1999a: 245) calls this 'a true interdisciplinarity', in which mapping was to be the process of creating abstract concepts and images that travel.

As Godlweska (1999a: 252) shows, von Humboldt experimented with graphical representation in an attempt to adapt representational topographic methods to analytical, constructive images: 'Humboldt was seeking a more analytical spatial language that would allow the almost intuitive transfer of understanding from one graphic genre to another and from one specialist body of knowledge to another.' The result was a proliferation of 'isolines, distribution maps, flow maps, a map of error, proportional squares, something he called "pasigraphy", and a multidimensional pictorial graph' (p. 252). This was a directed experimentation aimed not only at the mapping of South America for Europeans, but the creation of an abstract analytical science; an Enlightenment project of science and liberty.

The repressive effects of such Enlightenment scientific principles and practices in the periphery of an expanding capitalist economy is, by now, well known. But Pratt closes the chapter on von Humboldt with a question of vital importance to any critical understanding of mapping practice:

> What hand did Humboldt's interlocutors have directly or otherwise, in the European re-invention of their continent? To what extent was Humboldt a transculturator, transporting to Europe knowledges of American origin; producing European knowledges infiltrated by non-European ones? To what extent, within relations of colonial subordination, did Americans inscribe themselves on him, as well as he on America?
> (Pratt 1992: 135).

Here Pratt (1992: 138) highlights the heteroglossic nature of all mapping, and forces us to ask about the forms and ideas that emanated from the 'contact zone' back into Europe through these various transculturated images. Particularly striking in Pratt's deconstruction of von Humboldt's geographies of Spanish America is their utter contingency. As von Humboldt renders the Americas in terms of sublime spectacle for bourgeois European consumers, as resource for its industries, and as wild nature in need of civilizing powers for its expansionist forces, the rich and deep cultural practices of indigenous and Spanish colonial peoples were largely erased. In the practice of rendering such a spectacle, however, von Humboldt turns out to have been a rich transculturator of local knowledges, inserting traces of the Other at the heart of the construction of European identity.

This is, I think, an important lesson for our thinking about maps and power, about how cartographic and geographic concepts and practices become sedimented, how they become produced as facts and as the 'Real'. Von Humboldt as the cartographer of Spanish America for Europe, and as transculturator, is the vehicle for both the mapping onto Spanish America of 'Imperial Eyes' and a distinctly European modernity (with all its untoward effects in the nineteenth and twentieth centuries) *and* a channeller of distinctly South American knowledges into the heart of that hegemony. As Derrida (1998) has suggested recently, language – even imperial language – in this sense is never univocal, but always a dialogue of histories, cultures and places, shaped by those it possesses as well as those that seek to control it. Such a history of mapping is only now beginning to be written.

In this sense, it is also clearer why Napoleon never did appreciate the political possibilities of von Humboldt's science and his understanding of planetary consciousness. Upon meeting von Humboldt, Napoleon curtly dismissed his scholarly and scientific pursuits as of little interest to the states*man*, although they might be of interest to his wife who, he suggested, was also 'interested in botany'? (Napoleon's (only) words to Alexander von Humboldt (1805); quoted in Pratt 1992: 111). Von Humboldt's project contrasted starkly with the pragmatic politics of the arch-imperialist, Napoleon, for whom mapping was purely an instrumental and strategic tool of military power.

These two Euro-expansionist projects competed and complemented each other. Larry Wolff (1994: 145) has argued persuasively that from the eighteenth century European mapping projects were first and foremost a geopolitical and cultural exercise in the recapturing of what were referred to as 'these lost lands'.

> The map on the table had served as an invitation to conquest at the beginning of the eighteenth century; the map on the medallion, at the end of the century, could be carried home to Paris, London and Vienna ... It was a token of the fact that the lost lands of Eastern Europe had been discovered, mapped, travelled, studied, and stamped, according to the enlightened standards of western Europe.

The representational economies of South America and Eastern Europe were quite distinct, deploying the same graphical tools and analytics in distinctively different registers and tones. While for Napoleon mapping was a strategic tool for the deployment of imperial power over the lost territories to the East, for von Humboldt, mapping was part of a broader scientific deployment of abstract categories and images, whose effects were – in the end – far more deeply rooted in Europe and elsewhere. Von Humboldt's abstract diagrams and maps figured the 'planet' but erased the peoples and landscapes of the Americas. But it did so paradoxically by drawing on indigenous knowledge systems, local community networks, and diverse concepts and experiences.

How then were diverse regions territorialized in terms of universal and univocal narratives of modern nationalism? How were the religious, regional, linguistic and ethnic differences 'rendered spurious, reactionary, and vestigial?' (Krishna 1996: 82). And how were discourses of nationalism, sovereignty and citizenship mobilized to produce the body politic we know of as the modern nation-state? That is, as Thongchai (1994: x) asked, how is it that nationhood and territory have been 'arbitrarily and artificially created by a very well known science – namely, geography and its prime technology of knowing, mapping – through various moments of confrontation and displacement of discourses'? How was the geo-body of the territorially bounded nation-state discursively created?

Like Edney (1997), Krishna (1996: 82) asks these questions from the perspective of contemporary India and locates cartography at the heart of the constitution of the spatiality and body politic of the state:

> By *cartography* I mean more than the technical and scientific mapping of the country. I use the term to refer to representational practices that in various ways have attempted to inscribe something called India and endow that entity with a content, a history, a meaning, and trajectory. Under such a definition, cartography becomes nothing less than the social and political production of rationality itself.

It is in this 'social and political production of rationality itself' that maps, cartography, and broader geographies also played important roles in decoding and recoding the European *socius*; Orientalism was also a coding of European, metropolitan subjectivity. The role of maps in the production of subjects as citizens, consumers, and loyal 'subjects' occurred throughout the new territories of Europe's empires, but it also occurred in the hearths of industrial capitalism. The colonizing of indigenous knowledge and the uses to which maps and science were put may have deepened the crisis of representation in the Americas and contributed to the shaping of a particular form of political modernity in Asia, but the technologies of nature and society they ushered in – along with Napoleon's strategic conception of mapping – became an important part of the technologies of rational planning of the social body at 'home'. It is to this social body that we now turn.

7   Commodity and control
Technologies of the social body

We should be led too far, if we developed our belief as to the transformations to be wrought by this greatest of human triumphs over earthly conditions, the divorce of form from substance. Let our readers fill out a blank
check on the future as they like, – we will give our endorsement to their
imaginations before-hand. We are looking into stereoscopes as pretty toys,
and wondering over the photograph as a charming novelty; but before
another generation has passed away, it will be recognized that a new epoch
in the history of human progress dates from the time when He who
– never but created light
Dwelt from eternity
Took a pencil of fire from the hand of the 'angel standing in the sun', and
placed it in the hands of a mortal.
(Oliver Wendell Holmes, 'The stereoscope and the stereograph')[1]

COMMODITY, FETISH, ABJECTION

In 1996 the first issue of *Mercator's World* was published. *Mercator's
World* is a glossy, bimonthly magazine of map collecting, a kind of Jay
Peterman for map collectors. Peterman's rugged traveller extolled the
virtues of individualism in exploring the world's 'wild' places, taming
them, in order to be able to bring to middle-class consumers exoticized
commodities from around the world. *Mercator's World* very first issue
began with the following editor's introduction:

As a young man of 16, I left home and started wandering the world –
from California to London's Soho, to the august environs of Heidel-
berg. On to the Straits of Gibraltar, then East Africa, back to California, south to Australia, north to Alaska and the Yukon Territory, and
finally after four years, back home to Los Angeles. What a wander – a
great trek with many adventures along the way. The thread that held
many of my voyages together was the maps that guided me. The
intrigue of my travels – that first trip and hundreds since – has
somehow inextricably woven itself into the very fabric of maps and

they have remained a lifelong fascination. And as publishing has been my profession over the past 25 years, it's inevitable I suppose, that the two passions would eventually meld.

And so *Mercator's World* is born. We hope to bring to the reader the beauty and art of the great map-makers – cartographers of yesterday melded the mystery of undiscovered worlds with the information gathered by the explorers, and map-makers of today face the exact same challenge.

I have met many of the major map dealers and some of the great map collectors and the passion experienced is altogether overwhelming – to hold a guide that a renaissance prince used to navigate around the world in search of trade is mind boggling. To know that there is only one of three of something in the entire world and that you have access to it in some way is in itself exciting.

(Aster 1996: 7)

In a 1994 essay entitled 'The system of collecting' Jean Baudrillard considered the ways in which the object – the loved object – functioned as part of a broader economy of identity formation. It is in the context of collected objects that self-understanding emerges in a world of private property. In such a world, it is the cretin who collects nothing. Collecting 'consists in the criss-crossing of categorical boundaries, the revelling in the jarring juxtapositions characteristic of the postcard album' (Schor 1994: 258). In a sense, *Mercator's World*, epitomizes this system of collecting commodities, a system with direct roots in the natural inheritance of von Humboldt's planetary and bourgeois sensibilities.

North, south, east, or west. Brown skin, yellow skin, red, or white. Rich, poor. Laborer, president, king. Bedouin tribesman. Sunshine, snow, hurricane, volcanic activity. Beaches, palaces, grass huts. Politics, war, social history. Holidays and humor. Street scenes and pastures. And always, always more.

(Smith 1989: 137)

Such notions of collecting and their underlying epistemology were at the heart of J.K. Wright's view of science. In his essay *'Terrae incognitae'* Wright was concerned with the closing of geographical categories wrought by totalitarianism and by the emergence of a parallel instrumentalism in the social and geographical sciences (see also 'Map makers are human' and 'Human nature in science'). Wright (1942: 83) urged geographers to be open to 'geosophy', which he defined as

the study of geographical knowledge from any or all points of view ... Taking into account the whole peripheral realm, it covers the geographical ideas, both true and false, of all manner of people – not only geographers, but farmers and fishermen, business executives and poets, novelists and painters, Bedouins and Hottentots.

His focus on the geosophic imagination and the subjective in maps have long resonated with geographers. Maps have fulfilled the 'collector's' fantasy, in part because of the intrinsic technical and aesthetic merits of the craft map, in part because of the worlds of mystery, novelty, and alterity maps opened to the adventurous (especially young boys), in part because of the magical acts of transmutation involved in map-making and map-reading. Professional geographers have long attested to the aesthetic and mysterious importance of their childhood encounter with maps. This was perhaps never more true than among the generation of British and American geographers who lived through the Second World War. Thus, in two volumes (*Conversations with Geographers* (Browning 1982) and *Geographical Voices* (Gould and Pitts 2002)) many of the individuals begin their autobiographical reflections with their encounters with, and love of, maps. Many of these admissions have a 'confessional' form and, I suspect, this is because each of these individuals was well aware of the complexity of an aesthetic that conjoins weekend pleasures with journeys of exploration, local pathways with the exploration and colonization of new territories and peoples, pictorial pleasures with the greater glory of the nation-state, and the joys of owning a map with the private property claims of its leading citizens.[2]

It is to these processes of subject formation and subjection that emerge in this complex aesthetic of map use and its associated social practices that I now turn. There are, I think, two important ways in which this love of maps and the 'geosophic' aesthetic played out historically. Here I focus on one of these forms; the commodification of culture and the ways in which such notions of 'local knowledge' extended the economy of display. In the final chapter, I return to the second of these forms and to the progressive moment in Wright's claims. Here I begin with the question of commodities and cultures of display. I then turn to the ways in which the map emerged as a diagnostic tool for social analysis, and in the process recoded social life and reworked the spaces of the city. The 'cleansing' of the moral and physical infrastructure of the city as a social project for diagnosing and controlling the masses was achieved in part by the rationalizing and mapping of space. Planetary consciousness, abstractive science and the moral topographies of modernity combined with an aesthetic sensibility to more deeply commodify the world-as-picture and produce the world for exhibit, display and collecting. Epistemology, political economy and morality were being reshaped. In this process, the map played a fundamental role in shaping modern societies everywhere, both in the hearths of industrial capitalism and the hearts of Empire, as well as in the farther reaches of modernist economic and political desire.

MAPPING AND THE PRODUCTION OF SOCIAL IDENTITIES

For Henri Lefebvre (1991), there was a fundamental distinction to be drawn between the structuring of everyday life by *representations of spaces* that involved conceived, planned, rationalized spaces of social production and *representational spaces* of everyday life. The 'degradation of space' and the progressive commodification of everyday life is achieved, in part, by representations of space such as maps, architectural plans, and other techniques of spatial rationalization.

Spatial fixes and individual citizen consumers must, of course, be produced and reproduced, and in this section I focus on the ways in which maps and mapping have functioned to produce these spaces and kinds of individual. That is, the cartographic practices of high modernity contributed in various ways to the coding of the *socius* and the deterritorializing and reterritorializing of nineteenth-century and hence contemporary social spaces. In particular, I want to focus here on some specific ways in which modern social identities were constituted in and through the practices of nineteenth-century mapping. In so doing, they produced new subjects, new identities, and new ways of understanding information. In this sense, Oliver Wendell Holmes was absolutely correct when, in the quotation with which this chapter began, he suggested that 'before another generation has passed away, it will be recognized that a new epoch in the history of human progress' was in the making.

Ola Soderstrom (1996) has 'drilled' what he called three boreholes at crucial historical junctures into the history of technologies of visualization and urban planning to illustrate the ways in which urban space was revisualized through the techniques and practices of mapping. The metaphor of boreholes signifies Soderstrom's commitment to writing episodic forms of historiography, to the ways in which technologies represent – not progressive, linear evolutions – but 'a series of essential bifurcations in the modalities of visualization of urban space' (Soderstrom 1996: 251). The denaturalizing of their social histories aims to contribute to the broader project of 'calling into question the transparency of representation' (p. 249) and replacing it with detailed genealogies of the origins of 'claims to the totalizing, disembodied gaze upon the object of analysis', a gaze which disqualifies 'the richness of partial, situated points of view' (p. 250) and creates in its place 'a slippage of individuals towards objects' (p. 274). The three episodic histories (the ichnographic plan, the master plan and zoning plan, and social cartography) 'synthesize the city in terms of material objects, or individuals who are treated as objects, that is, reduced to social types, operators of functions (living, working, travelling, recreating) or of standard needs (norms of comfort, of noise, of household goods)' (Soderstrom 1996: 274–5). I draw on Soderstrom here to illustrate the specific ways in which what he calls the zenithal gaze emerged within the apparatuses of the state, giving rise to new forms of social cartography and thoroughly spatialized conceptions of social identity.

Borehole 1: Planimetry and the ichnographic plan

As we have already seen (Chapter 4), the reinvention of linear perspective signalled a change in the ways in which landscape and urban form were represented. But perspective was limited by its dependence on the fixed gaze. Urban planners needed a system of representation that allowed for the rationalization of urban space and in the mid-fifteenth century this was provided by Leon Battista Alberti. In *De re aedificatoria*, *De pictura*, *Ludi matematica* and especially *Desriptio Urbis Romae*, Alberti presented a rational method for mapping the city that could be followed by 'anybody of average intelligence', did not depend on exceptional artistic talent (thus making the task of mapping a technical one that could be carried out by anyone or any group of trained 'planners'), and did not depend on being embedded within an explanatory narrative. The result was, as Soderstrom (1996: 258) argues, that: 'Visualization is self-sufficient, containing within itself the terms of its own signification.' Alberti had described how to develop and use a geometrical plan and this quickly became the basis of modern planimetry and what we now commonly take to be a city plan (Soderstrom 1996: 256). From Leonardo da Vinci's first truly ichnographic plan (his 1503 plan of Imola), the techniques of this new system for visualizing urban space diffused quickly. No longer did the representation of urban space depend upon the fixed view of a single observer. Instead it had become 'the net product of multiple points of view' (p. 258); a totalizing, stabilizing and scaling of urban space (Figure 7.1).

Figure 7.1 Bird's-eye view of Phoenix (Library of Congress, Washington, DC)

Borehole 2: The master plan

The success and speed of adoption of the ichnographic plan quickly established this form of urban spatial planning as the norm for military, planning and administrative purposes. Bird's-eye views remained popular with the public (Schein 1993), continuing to operate alongside the more abstract, geometrical plans. But, Soderstrom (1996: 26) suggests, the expansion of uses for urban plans paradoxically produced a shift in the gaze of the observer from horizontal–oblique views to the more unusual vertical. This required a general retraining of the scopic regimes of naturalized perspective and descriptive representations, naturalizing geometrical plans and God's-eye views of the city (the zenithal gaze).

> It is through the agency of the zenithal gaze, therefore, that one can represent a town in the form of zones characterized by standards of living, classes of population, differing crime rates. In other words, the geometrical plan is a prerequisite of the thematic urban cartography which would develop towards the end of the nineteenth century ... The shift to the geometrical plan for this reason represents a crucial step along the road leading to the development of Victorian social statistics, since the urban zones constituted by the zenithal point of view could then form the subject of procedures of census, measure, and comparison.

The emergence of the master plan to ensure effective administration and the proper functioning of the city was stimulated by the very concerns of public administration we discussed above (rapid population increase, increasing urban residential densities, social tensions, and worsening sanitary and health conditions). In mid- to late nineteenth-century Germany, these concerns spawned a strong interest in urban planning. Soderstrom focuses on the work of one such urban planner from Karlsruhe, Reinhard Baumeister. In 1876 Baumeister published the first high-circulation urban-planning manual *Stadterweiterungen in technischer, baupolizeilicher und wirtschaftlicher Beziehung*, in which he argued for the use of a master plan 'ensuring the stability and proper functioning of a city conceived of in terms of a living organism in order to deal with the problems it faced: overpopulation in certain districts, problems of traffic and hygiene, social unrest' (Soderstrom 1996: 262). The urban master plan was to monitor land values, maintain urban order, and allow investors to anticipate future development in the city, and soon the master plan incorporated zoning plans that functioned as spatial and temporal 'processing devices' 'transforming the city into a "logical and predictable structure"' (Soderstrom 1996: 263).

A visualized and mapped urban imagination diffused quickly in the nineteenth century as planners and administrators struggled to deal with the instabilities of new urban form and life. In these efforts, the visualizing of the city and the mapping of its elements became an obsession of urban planning. The very name 'planning' – the making of plans – contains within it this dual understanding of the function of urban administration. First, the rendering of the social and built environment in graphical form (the 'plan'), and second, the temporal extension of needs and desires from the 'plan' into a 'plan' for the future. Soderstrom (1996: 267) refers to Patrick Geddes, for example, as a 'visual obsessive' and 'inveterately oculocentric' in his efforts 'to scale the whole world down to graphic form'. Committed to the use of cartography and statistics, Geddes argued vigorously for social and geographical surveys as crucial elements in urban planning and, with almost Benjaminian devotion, his desire to create systematic graphical representations of the city

> culminated in the *Index Museum* project, an enormous graphic encyclopaedia, the ultimate and all-embracing form of the total museum which he tried to make a reality, without much success, by making the 1900 Paris World Fair permanent.

Borehole 3: Urban social cartography, or 'The visual order of the civil servants' (Soderstrom 1996: 261)

As Paul Rabinow (1989: 18) has argued: 'All three terms – Man, observation, and society – were in the process of changing meanings in the long threshold between the classical age and modern times.' This was a time when the 'era of Man' emerged, before which 'the world, its order and human beings existed, but man did not' (Foucault 1973: 322). At this time, 'Man appears as an object of knowledge and as a subject that knows' (Foucault 1973: 319). With what Rabinow (1989: 30) calls the 'watershed event' of the cholera epidemic of 1832, 'new scientific discourses, new administrative practices, and new conceptions of social order, usher [...] in a long period of experimentation with spatial/scientific/social technologies.'

The result, Rabinow (1989: 39) suggests, was that 'The apparatus of finely grained observation of the social body – supervised by physicians, aided by architects, and backed by the police – in the service of the health of the population and the general good, had a long career ahead.' One aspect of these technologies of the social body and public norms was the role of information about the individual. Along with social statistics, architecture and city planning, the map played a significant role.

The first pocket-sized map designed to be used on the street was produced in London in 1854, *Collins' Illustrated Atlas*. The atlas created a new kind of visibility in the city and, correspondingly, new behaviours by both the public and the state. Nowhere was this more clear than in Dr John Snow's mapping of the 1854 cholera epidemic of London. The 500 recorded deaths from cholera were each represented by a skull on a map of the Golden Square neighbourhood of London, and the eleven water pumps used by local residents were also mapped. Snow used the map to make his now famous observation that cholera occurred almost entirely among those who lived near to the Broad Street water pump and from which they obtained their drinking water. With the removal of the pump handle, the epidemic ended and the era of public health and epidemiology began. Snow's work and its effects spatialized the newly emerging institutions of public health (following the Public Health Act of 1833) and extended the field of the state apparatuses by framing the 'public' and the 'population' as spatialized objects of observation, mapping and administration. The new mapping and measuring technologies became standard tools for solving problems of administering the new public sphere.

One by one, the institutions of the police, public health, architecture and urban planning were all rendered possible by the technologies of political statistics and mapping (Rabinow 1989: 74). Between 1886 and 1903 Charles Booth produced maps covering many aspects of urban space and society (Figure 7.2). Such social mappings were a crucial tool for summarizing and presenting the results of social surveys throughout the city: 'This objective gaze on the London slums was highly efficient, since it led to a naturalization of the moralizing discourse on poverty and was at the origin, as he wished it to be, of "scientific" social policies' (Soderstrom 1996: 269).

Jacob Riis's *How the Other Half Lives* (1890) was indicative of a parallel concern for scripting social life in the interests of public health and safety in cities of the United States. Focused on New York City and based on detailed interviews, social statistics and photographs, Riis's work was part of a series of wider-ranging efforts to document and categorize the poor in the city. Such projects of public administration and social and moral policing, particularly aimed at the 'unruly' (or the threat they were perceived to pose) were repeated in city after city in the industrial countries of the North and in the settler economies of the South (on the former see Dorn 2002, on the latter see Popke 1999).

Mapping, social statistics, public health, moral education and the institutions of urban planning and administration were from the beginning thoroughly spatialized practices of identity construction and social engineering. Mapping and statistics made citizens visible in particular ways, rendering them subject to public administration, and it provided useful accounting tools for new practices, such as insurance. For example, Sanborn maps in the US recalibrated the urban scale in fundamentally new ways (primarily in terms of ownership and function), introducing a scale of detail and precision that had not before been mobilized to define urban space (Figure 7.3). These practices of mapping and remapping literally opened new spaces of social life as they coded the places within a rubric of territorialized state administration and commodified property relations. In the process, the individual was 'fabricated by this scientific technology of power that I have called "discipline"' (Foucault 1977: 194).

Figure 7.2 Mapping the social life of London. *Map Descriptive of London Poverty,*
    1898–9 by Charles Booth

Foucault (1977: 194) goes on to argue:

> We must cease once and for all to describe the effects of power
> in negative terms: it 'excludes', it 'represses', it 'censors', it 'abstracts',
> it 'masks', it 'conceals'. In fact, power produces, it produces
> reality; it produces domains of objects and rituals of truth. The indi-
> vidual and the knowledge that may be gained of him belong to this
> production.

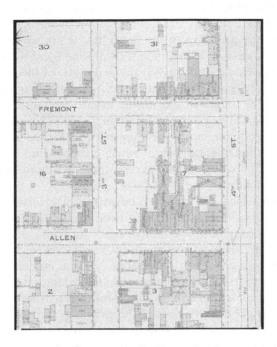

Figure 7.3 Fire Insurance Map, Tombstone, Arizona (New York: Sanborn
Map and Publishing Company, 1886) (Library of Congress, Washington DC)

And in this production, graphical representation has played a central role.
Again, as Soderstrom (1996: 274–5) has shown so well:

> The principal family resemblance between the ichnographic plan, the
> master plan, the zoning plan, and social cartography is that they syn-
> thesize the city in terms of material objects, or individuals who are
> treated as objects, that is, reduced to social types, operators of func-
> tions (living, working, travelling, recreating) or of standard needs
> (norms of comfort, of noise, of household goods).

MIRROR WORLDS AND CARTOGRAPHIC
REPRESENTATION

In the nineteenth century, the increasing deployment of maps as part of a
broad social project of urban and social rationalization was paralleled by
the use of cartographic rationality and its representational logics in new
forms of commodification. Marx recognized these more generally as the
driving force of capitalist societies: the constant extension of the commodity form to ever wider domains of everyday life and the fetishizing of more

and more objects for circulation as commodities. Capitalism is, as Deleuze and Guattari (1983: 245–6) have shown, 'the [relative] limit of all societies, in so far as it brings about the decoding of the flows that the other social formations coded and overcoded ... It axiomatizes with one hand what it decodes with the other.' For Baudrillard (1988) the axiomatic of the commodity form is the necessary and appropriate condition for all objects within a capitalist society. The nature of citizenship in late capitalism was to act as an individual, self-satisfying consumer. Claims about the lack of moral fibre, social anomie and community disintegration deflected attention away from an equally fundamental way in which citizens were being scripted, as consumers. The individual citizen as consumer, in other words, is one spatial fix for late capitalism.

Mercator's World (with which this chapter began) straightforwardly links the map with the processes of commodification and cultural decoding and recoding. But in its collecting of images across space and time, *Mercator's World* also links the hyper-commercialism of twentieth-century *fin de siècle* with the fascination with collecting that permeated the nineteenth-century *fin de siècle*. Here the museum, panorama and exhibition functioned alongside the map to constitute a particular form of experience and understanding of modernity. The fetishizing of the map was part of this wider fetishizing of vision and the construction of a particular visual optic and scopic regime that came into play in the nineteenth century. It is, I think, one that continues to inform much contemporary understanding of the geographical and cartographical imagination.

As Thomsen (1994: 96) has shown, 'the 19th century loved inventions and technical innovations. Growing industrialization produced in the leading European countries a need for new media of pictorial representation.' Technical innovations abounded. Many have become such commonplace items that we rarely even notice them today. Yet the resultant cultures of display transformed both public space and nature. For example:

> On June 19, 1787, Robert Barker, an Irish-born artist working in Edinburgh, was awarded a British patent for his invention of a type of large-scale, 360-degree painting that he called 'nature at a glance.' When he introduced this device in London in 1792, it acquired the name – panorama – that stayed with it. By the mid-1800s every big European and American city boasted permanent buildings for the display of regularly changing panoramas. Viewers, after paying an admission fee, moved through a dark passageway and arrived at a raised platform in the middle of a cylindrical picture – sometimes reaching up to 22 yards in height and extending to 153 yards in circumference – that offered a complete circular view of a landscape, a cityscape or a famous battle scene. To increase the illusion, the upper edge of the canvas panels was usually hidden from sight, and the space between the viewing platform

and the painted imagery was often filled with real objects, such as shrubbery. Surrounded by the painted image, spectators could immerse themselves in a vast detailed view of some distant place or event. A French newspaper account of 1807 enthused, 'After five minutes, you no longer see "painting"; nature herself is before you.'

(Rice 1993: 69)

By the mid-1850s 'panorama mania' had become a full-blown form of visual entertainment in major European cities, creating a spectacle of images that depended in part on cartographies of various kinds. Significantly, 'panorama' is a neologism from the Greek meaning 'all seeing' (Miller 1996: 35), and was originally coined in 1792 in a notice in the London *Times* announcing the appearance of Barker's Edinburgh panorama (Miller 1996: 39). During the nineteenth century, the notion of panorama expanded to include any picture, large or small, that provided a 'sweeping overall view of its subject' (Rice 1993: 70), including

> maps and diagrams, bird's-eye topographic views, painted and photographed images encompassing a broad range of vision, elongated-format images describing a site or situation, magic-lantern projections and a variety of visual toys. There were even such oddities as an embalmed pigeon with a small camera strapped around its stomach, trained while alive to take photographic 'bird's-eye views'.

The broadening of content was matched by a proliferation of forms: to the more standard representational forms, such as maps, photographs, and postcards, were added cosmoramas, dioramas, neoramas, cycloramas, moving panoramas and pleoramas (Miller 1996: 36, 42), each representing and rendering the exotic and distant as a commodity for consumption.

This nineteenth-century fascination with visual representations also extended to the emerging professional fields of surveying. In France, Laussedat – dissatisfied with free-hand perspectives used in field mapping – was the first to incorporate *the camera lucida* to produce perspective drawings. The procedure known as *iconométrie* (image or likeness) was well received not only because of its accuracy (up to a distance of 1.5 km when using a telescope), but also because of its greater flexibility over the *camera obscura* in the outdoors. Convinced that the future of mapping could be based on the use of photographs, Laussedat completed the first complete topographical mapping of the Village of Buc near Versailles (Figure 7.4). The map, the field processes used, and the speed and accuracy that resulted catapulted photogrammetry and Laussedat into national and international prominence. Laussedat became professor of geodesy at the Ecole Polytechnique in Paris in 1856 and director of the Conservatoire National des Arts et Métiers in 1881. He was elected to the French Academy of Sciences in 1894 (Blachut and Burkhardt 1989: 89). The map

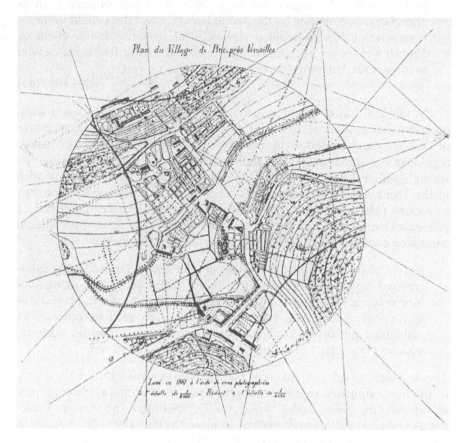

Figure 7.4 Award-winning map: map of the village of Buc, Versailles, at the scale of
1:2,000 produced from photographs in 1861. The map won a gold medal
in 1863 in Madrid (Blachut and Burkhard, 1989, with permission, American Society for Photogrammetry and Remote Sensing: The Imaging and
Geospatial Information Society)

was awarded a gold medal in 1863 in Madrid and later travelled to the
exhibitions in Paris and Chicago.

By mid-century, the panorama had became a popular cultural icon, spectacle and commercial opportunity. 'Panoramic consciousness' had become
such a 'public obsession' that by 1848 one panorama of the Hudson River
was reported to have been 12,000 feet in length, and between the middle
and end of the century roughly 300 giant panoramic productions had been
seen by an estimated 100 million people (Miller 1996: 36). There had
emerged what Miller (1996: 35) describes as 'an international hunger for
physically, geographically, and historically extended vision'.

Like present-day NASA imagery, public excitement for the map and the panorama was used to garner further support for new technologies such as photogrammetry, hot-air balloons, train travel, and high-rise buildings, as well as to boost the image of locations and cities depicted (see Schein 1993 on the parallel use of bird's-eye view imagery in nineteenth-century America). In the second half of the nineteenth century, photography merged quickly with the panorama, and stereoscopic panoramas of geographical images from around the world became popular. The Kaiser-Panorama of Berlin in the 1880s typified these emerging geographical imaginaries, combining popular spectacle, bird's-eye vision and the celebration of new possibilities in mass transport: Paris, Berlin, through the Panama Canal, into the Andes mountains, along the Trans-Siberian railroad, and in hot-air balloon flights (Rice 1993: 71). Such panoramas catered to a growing public desire for spectacle, particularly among the burgeoning middle class, whose demand for recreation, travel, and the consumption of images fuelled not only the panorama industry, but also tourist resort development (seaside resorts in particular),[3] the postcard industry (see Schor 1994), and a host of visual toys (Figure 7.5). In short, the nineteenth century produced a rich representational economy of mappings involving complex technologies of capture, rendering and presentation (literally so in the case of museum displays and life studies). Through them a particular world-view was being structured: a world-as-picture, -as-exhibition, -as-museum, and -as-miniature. But this rationalized scaling of cultural and natural alterities was also a form

Figure 7.5 The diorama and mirror worlds of *fin de siècle* Paris

of subjection in which spectacle and commodity mediated the production of a subject whose very identity and consumption patterns were *from the beginning spatialized, globalized and gridded.*

The production of such modern subjects is illustrated nowhere better than in Walter Benjamin's *Passagen-Werk*. This project was carried out in Paris up to and during the early years of the Second World War with the explicit goal of investigating the representational economies and cultural transformations at work in nineteenth-century Paris at a time of major capitalist restructuring. In that project Benjamin was concerned with the emergence of notions of history that were dominated by teleologies of progress; linear histories of the natural unfolding of white, male, European history; histories that naturalized difference and ignored the violence and erasures that typified commodity culture.

In turning to Paris, Benjamin insisted that what was new at the time was not the urban brilliance and luxury of the city, but secular public access to them (Susan Buck-Morss 1989: 81), a form of secular public access that Matless (1999) has described as fuelling the demand for topographic maps in Britain during the same period. Paris was, in this sense, a 'looking-glass city' and a mirror city that dazzled the crowds, reflecting images of new consumer goods and consumers, but it did so by 'keeping the class relations of production virtually invisible on the looking glass's other side'. This was an emerging economy of spectacle and display, what Benjamin called '"phantasmagoria" – a magic lantern show of optical illusions, rapidly changing size and blending into one another' (Buck-Morss 1989: 81). In this system everything desirable came to be transformed into fetishized images of commodities-on-display, and when newness became a fetish 'history itself became a manifestation of the commodity form'.

Benjamin sought to unmask this fetishized mirror world of end-of-century Paris by describing what he called the 'ur-forms of the phantasmagoria of progress'. Four such ur-forms are of direct interest to our present discussion: the panorama, the arcade, the world exhibition and the plate-glass shop window. Each represents elements of the informational transition that was occurring in the late nineteenth century as western capitalist economies internationalized and new global imperial geographies were built. As we have just seen, the panorama was a new technology of visual representation organized and moved around different cities to present spectacles of one form or another to eager middle-class consumers. The panoramas provided sweeping views that rolled by the viewer at varying speeds, giving the impression of movement through the world at accelerated speed (Buck-Morss 1989: 82). Panoramas were a common feature of the new commercial arcades springing up throughout the city ('the original temple of commodity capitalism'), and it was in the arcades that the increasingly globalized flow of images and commodities came together (Figure 7.6). The arcades were the precursors of the department store and, in more contemporary form, one might see in the con-

Figure 7.6 Technology, globality, commodity, anonymous. Air Bleu, France-
Afrique du Nord, Passagers, Messageries, Les Avions Bleus poster,
1938. Cassidy's Empire Chocolates, poster, undated. Black and
white: (a) 'On display' (b) 'A glass arcade' (Catalogue of the
Crystal Palace Exhibition, London, 1851)

joining of the panorama and the arcade precursors for the digital world of the
Internet and online shopping.

The culmination of the panorama and arcade experience was the emergence of the great world exhibitions, the first being in London in 1851 – a
mirror world of a different kind; a 'Crystal Palace'. It was in these great international exhibitions and fairs that the 'pleasure industry' had its origin and it
is they that 'refined and multiplied the varieties of reactive behaviour of the
masses. It thereby prepares the masses for adapting to advertisements.' The
advertising industry and world exhibitions shaped a mass public that was at
one and the same time individualized, nationalized and globalized.

The exhibitions and arcades incorporated another technology that
became fundamental to a modernist sensibility: the large plate-glass
window. This gave to sellers the ability to display goods but prevented

consumers from touching. Pleasure was now to be derived from the visual spectacle alone. The representation of far-away places and possible ways of life came, in itself, to be a source of pleasure, as was the broadening experience and promise of movement, global reach, and speed. Exhibitions and arcades were, then, for Benjamin the source of a broader phantasmagorical politics: 'a promise of social progress for the masses without revolution' (Buck-Morss 1989: 86). 'Each successive exhibition was called upon to give visible "proof" of historical progress toward the realization of these utopian goals, by being more monumental, more spectacular than the last' (Buck-Morss 1989: 87), and each show-cased the technologies that enabled the movement of goods around the globe. Speed, information, and access came to symbolize progress, and the globe came to symbolize this particular form of modernity's promise.

Like the aerial navigators in Jules Verne's *Robur-le-Conquérant* (Paris 1886, London 1887), surveying the terrestrial scene had now come to embody not only the visual giddiness of elevation and the ability to survey the earth beneath, but it also enabled the consumption of such images at ever increasing speed. The cartographic eye had become a central element of cultural consumption and technological imagination. The future was to be richly and deeply geographical (see Mattelart 1999: 80–2).

For Benjamin, the mythic history of progress embedded in these cultural representations was so generalized that the possibilities for dislodging its hold on the masses was extremely limited. He resolved his dilemma by a search for 'counter-images' and through these small, discarded objects (the trash of history) he sought to illustrate a different conception of history from which all traces of Progress and development were eradicated.[4] Paul Klee's painting, *Angelus Novus*, provided a map for this vision of history, which stood in marked contrast to the futurist myth of historical progress which could only be sustained by forgetting its past (Buck-Morss 1989: 95):

> There is a picture by Klee called 'Angelus Novus.' An angel is presented in it who looks as if he were about to move away from something at which he is staring. His eyes are wide open, mouth agape, wings spread. The angel of history must look like that. His face is turned toward the past. Where a chain of events appear to us, he sees one single catastrophe which relentlessly piles up wreckage upon wreckage, and hurls them before his feet ... The storm [from Paradise] drives him irresistibly into the future to which his back is turned, while the pile of debris before him grows toward the sky. That which we call progress is this storm.

At the heart of mythic notions of history are a series of metaphors and images. These Benjamin called 'wish-images', and they remain at the core of modernist and liberal conceptions of history as progress:

These images are wish images, and in them the collective attempts to transcend as well as to illumine the incompleteness of the social order of production. There also merges in these wish images a positive striving to set themselves off from the outdated – that means, however, the most recent past. These tendencies turn the image fantasy, that maintains its impulse from the new, back to the ur-past. In the dream in which every epoch sees in images the epoch that follows, the latter appears wedded to elements of ur-history, that is, a classless society ... Out of it comes the images of utopia that have left their traces behind them in a thousand configurations of life from buildings to fashions.

(Benjamin quoted in Buck-Morss 1989: 114, 118)

In this new world of images, commodity fetishes and dream fetishes become indistinguishable. Food and other commodities drop magically onto the shelves of stores, and advertising and commerce come to be seen as the means of social progress (Figure 7.7). The democratization of culture is now seen to derive from the mass media, and they too become fetishes (Buck-Morss 1989: 120). But the commodification of public space and the emergence of representational economies was also a process of subjection: composing and recomposing individual and collective subjects – the mapping of a new cultural politics of industrial life.

Figure 7.7 'Human Happiness – Food for the Asking in the Fourierist Utopia'. Grandville, 1844

Part IV

Investing bodies in depth

In the long run, this panoptic power could do much to shine light into darkness and make the world a better place. In areas like resource management it will allow citizens and their governments to make decisions with more information and in a greater spirit of openness than ever before. People in the developing countries, where resource management matters so crucially and where information asymmetry between the governors and the governed is often the greatest, may stand to profit the most. If, that is, ways can be found to afford them access to the data. The private sector is providing the world with these new eyes as a business proposition; the vision required to put them at the service of those who need them will have to come from elsewhere.

(Morton, 'A launch for the little guy: Satellite technology can now help real people')

The formalization and diffusion of computer-generated imagery heralds the ubiquitous implantation of fabricated visual 'spaces' radically different from the mimetic capacities of film, photography, and television.

(Crary, *Techniques of the Observer*)

Just as water, gas, and electricity are brought into our houses from far off to satisfy our needs in response to a minimal effort, so we shall be supplied with visual or auditory images, which will appear and disappear at a simple movement of the hand, hardly more than a sign...

(Valéry, *Aethetics*)

For the Enlightenment, whatever does not confirm to the rule of computation and utility is suspect ... Enlightenment is totalitarian.

(Horkheimer and Adorno, *The Dialectic of Enlightenment*)

8 Cyber-empires and the new cultural politics of digital spaces

One of the foremost tasks of art has always been the creation of a demand which could be fully satisfied only later. The history of every art form shows critical epochs in which a certain art form aspires to effects which could be fully obtained only with a changed technical standard, that is to say, in a new art form.

(Benjamin, *Illuminations: Essays and Reflections*)

It is now fairly well established in critical studies (if not in broader usage) that the 'Cartesian/Cartographic Anxiety' – the epistemology of modernist, universalist inquiry – has been pretty much laid to rest as a foundation for science. In its place we have more nuanced and multiform understandings of cartographic practice in which the production of geographical images is understood to be a thoroughly social and political project. Maps no longer are seen to simply represent territory, but are understood as producing it; in important ways 'maps precede territory', they inscribe boundaries and construct objects that in turn become our realities. Far from being a mere representation of private property, cadastral mapping gave legal and material form to the new territories and landscapes of private property. Geomorphological mappings do not so much mirror the physical world, but create textual abstractions that name and give structural form to the myriad fluxes and flows; they inscribe form and process through graphical representation and circulate these in ways that render them real. Booth's maps of London, Riis's documentary photographs of New York, Sanborn insurance maps, and Laussedat's photogrammetric maps did not merely mirror the socio-spatial patterns of working-class neighbourhoods, the structure of urban poverty and the morphology of modernizing landscapes but produced them as spatialized social categories in which new ways of thinking and representing people and places came into being, categories that in turn have since shaped urban social research and policy.

As the mapping and remapping of economic, social and political life has continued, multiple communities of users with new capacities for action, new imagined communities, and new 'spaces' for individual and collective identity have emerged. In each case, new technologies of mapping and new uses for maps have accompanied the reworking and recoding of social life. From CAD–CAM design and production systems, geo-referenced database mapping for insurance, marketing and polling companies, sophisticated self-guided missile systems, to three-dimensional imaging systems of the internal organs of the human body, cybernetic mapping systems have been in*corp*orated – literally 'embodied' – into social life, consumers for these new products and practices have been painstakingly produced, and new mapping metaphors have been deployed to promote yet further commodification and penetration of everyday life (Figure 8.1).[1] Images of a whole earth, representations of relationships that transcend local, regional, or national identities, new notions of community that transcend parochial conceptions of locality and place, and new mediations of self and other (constituted through digital interfaces and new representational forms) all become realities through these mappings of nature, society and the body-subject.

As the new digital mappings wash across our world, perhaps we should ask about the worlds that are being produced in the digital transition of the third industrial revolution, the conceptions of history with which they work, and the forms of socio-political life to which they contribute.[2] This chapter focuses on these transformations of mapping capabilities, their impacts on social life and their broader implications for contemporary democratic practice. Specifically, it situates the tools and approaches of digital mapping within wider transformations of late capitalism. In the second half of the chapter, I turn to the productive and progressive possibilities of these mapping systems and the reworking of cultural codes they seem to perform. This involves the creation of new visual imaginaries, new conceptions of earth, new modalities of commodity and consumer, and new visions of what constitutes market, territory and empire. My focus here is thus a reading of a broader history of spaces in terms of digital mappings of nature and society in which I show some of the ways in which the digital transition is also a transition in the economy of discourses within which truth claims are produced (see Foucault 1979, Kellner 1990, and Lefebvre 1991).[3]

Paul Valéry's (1964) understanding of the coming of a new age of information and images with which this part begins, or what Gianni Vattimo (1992: 1) has called a society of generalized communication, points to important issues about how we think about the relationship between mapping and the spaces of social life. Valéry's vision of the future penetration of imaging systems into every nook and cranny of everyday life reflects the universalizing of the cartographic impulse.[4] In this view, the elaboration of virtual worlds and spatial images extends our own world and our thinking about that world in remarkable new ways, opening virtual spaces for 'real' social interaction, new communities of dialogue, and new interactive

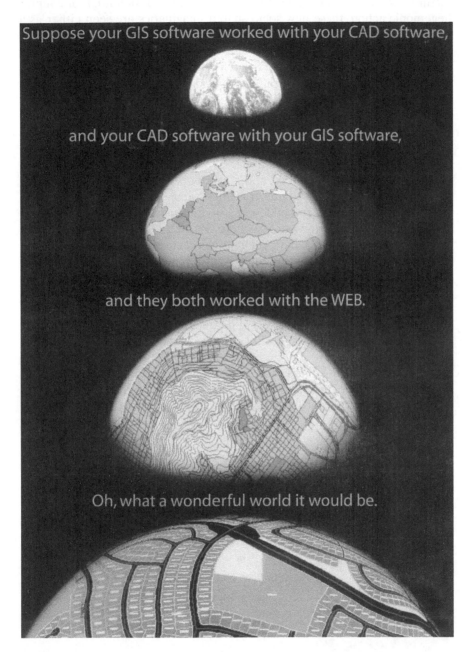

Figure 8.1 'What a wonderful world it would be'

settings for which we currently have only poor language and no architecture.[5] Whether as a resurgence of civic culture or as potential for counter
hegemonic action, these new socio-spatial imaginaries are seen either as a
stimulus to more effective running of accountable government or as a
potential liberator of socially and politically marginalized groups.

Even though the funding for research and development of the hardware
and software of modern mapping technologies has come primarily from
business, state and military sources, advocates of their progressive potentials
still argue that such geo-referenced information systems enable communities to make better decisions by providing access to more and better information. They provide more powerful tools for local planning agencies, exciting
possibilities for data coordination, access and exchange, and permit more
efficient allocation of resources, and a more open rational decision-making
process.[6] In this way, the 'digital transition' and the extension of mapping
practices are thoroughly embedded in a mythic history about the dissemination of democracy and the public sphere. Precisely how the views of social
progress that currently circulate in mapping practice and how they articulate
with a political and cultural economy of information, display and commodification remain open questions (Figure 8.2). If maps precede and produce
territories and social identities, what then are the objects and identities
being produced in the digital transition? And what forms of territorialization are at work in the new projects of digital mapping?

As Foucault (1980: 93) suggested in more general terms:

> in a society such as ours, but basically in any society, there are mani-
> fold relations of power which permeate, characterize and constitute
> the social body, and these relations of power cannot themselves be

Figure 8.2 'See the world in a whole new way'

established, consolidated nor implemented without the production, accumulation, circulation and functioning of a discourse. There can be no possible exercise of power without a certain economy of discourses of truth which operates through and on the basis of this association. We are subjected to the production of truth through power and we cannot exercise power except through the production of truth.

In what ways, then, do contemporary mapping techniques produce 'truth' and how in these ways do they 'exercise power'?

CYBER-EMPIRES?

The mappings of the digital transition have their own geographies. From one perspective these are part of a new Americanism, a thorough-going post-Fordism, and a resurgent geopolitics of global hegemony. They also have important implications for the ways in which notions of social progress are being written, global relations understood, and an American (and global) future is being mapped. By the late 1960s, prior to the widespread computerization of the 1970s, the primary and secondary information sectors of the US economy accounted for 46.2 per cent of national income.[7] The incorporation of cybernetic systems into the spheres of production and consumption since the 1970s has been exponential. Over the past two decades, technical change has transformed the scope and influence exercised over social life by the use of computerized data handling, imaging and mapping technologies. Particularly in the 1990s, there was massive reinvestment in the computerization of many aspects of economic, political and social life, accounting for a large part of the economic growth in northern industrial countries during that decade. One consequence has been that the top ten national information economies currently account for about 80 per cent of the global Information Communication Technology (ICT) market, whereas the bottom 10 represent a collective share of less than 1 per cent (WITSA 2000: 7), with North America comprising by far the largest regional market for ICT, spending $796 billion in 1999 alone.

Thus, the title of this chapter – 'Cyber-empires' – points to the social mobilization of these new mapping technologies and practices along with the broader technologies of information and communication of which they are a part (Figure 8.3). Beyond a political economy of technical change, a political technology of the social body and a corresponding regime of morality are emerging. Felix Guattari (1991: 18) called this 'the fabrication of new *assemblages* of enunciation, individual and collective' in which actors and scales of action are no longer only governments and nation-states, but complex assemblages that go well beyond the military industrial complex of the 1950s and 1960s. These emerging assemblages were described by Juergen Habermas in 1973 (255–6):

DESERT STORM'S SATELLITES OF WAR

Figure 8.3 'Desert Storm's satellites of war', cover *New Scientist* 27 July 1991, no. 1179 (*New Scientist*, Harcourt, London, with permission)

In industrially advanced society, research technology, production, and administration have coalesced into a system which cannot be surveyed as a whole, but in which they are functionally interdependent. This has literally become the basis of our life. We are related to it in a particular manner, at the same time intimate and yet estranged. On the one hand, we are bound externally to this basis by a network of organizations and a choice of consumer goods; on the other hand, this basis is shut off from our knowledge, and even more from our reflection.

In this regard, 'Cyber-empires' is also the name of a video game ('a challenging strategy game of world conquest') whose cyborgian characters emerge from the far side of the earth (laser weapons firing) to announce '[t]o the victor, the world. To the loser, the junk heap'. The techno-hype of the video-game producers illustrates sharply the roles played by the military, capital and the state in the development and deployment of new mapping technologies (Strategic Simulations 1992) (Figure 8.4).

Digital information systems, including geographical information systems and digital mapping, are at the heart of this new range of cultural and economic production. The social imaginaries and moral economies they produce have ever more profound effects the deeper and wider is their reach. At home and abroad, governments like the Reagan/Bush/Clinton/Bush Administrations script and inscribe new versions of state power, in which information-imaging systems facilitate military, political and economic goals simultaneously. Thus, the Gulf War, NATO's war against Serbia, and the war against al-Qaeda were the first GIS wars, although only the latest in a long line of wars using geographical

Figure 8.4 'Cyber Empires' (ad by Strategic Simulations Inc. in *Computer Game Review*, October 1992, vol. 2, issue 3) (with permission, UbiSoft Entertainment)

tools (see Clarke 1992). In the Gulf War, smart weaponry, GIS technology and telecommunications (including news organizations like CNN) were carefully orchestrated and coordinated – a 'kind of simulacra game in which the technology of entertainment television and the technology of mass destruction were deployed together as part of US military strategy to both deceive Iraqi military forces and to pre-empt/post-empt the formation of an oppositional public sphere' (Hanke 1992: 136).

I began writing this book at the time of the first bombing of Iraq, I completed the final draft at the time the people of Afghanistan experienced eight weeks of sustained and heavy bombardment from the most technically sophisticated army in the world. I complete final revisions as the current Bush administration seems determined to begin bombing Iraq again. This experience of sustained and repeated modern warfare has paralleled the writing of this book. More generally it has become a central motif of contemporary life. As Walter Benjamin reminded us, war and violence may indeed be the norm in modern society. Anyone growing up attentive to the constancy of hot and cold wars throughout the twentieth century must certainly agree. In these wars, increasingly advanced imaging and mapping practices have been deployed, particularly in the 'remote' campaigns and the broader strategic deployment of force in recent years. They have enhanced the power of the North and deepened the crisis of uneven development in the South, too often by 'primitive de-development' and 'collateral' destruction of civilian infrastructure and lives. As the comprador economy in the South has grown, mapping technologies have also been deployed by large corporations, multilateral lenders, and state bureaucracies to further natural resource exploitation and the provision and distribution of cheap primary products. As we try to understand the ways in which technology permeates the fabric of social life, creating new forms of individual and collective identity, it seems to me that – along with the many exciting new opportunities our imaging, information and mapping systems present for us – we must remember this broader geopolitical context and the ways in which these same mapping practices participate in fuelling new rounds of capital investment, creative destruction, uneven development, and indeed, at times, the ending of life, wrenching it from its moorings, and destroying it piece by piece, limb by limb.[8]

As the image from *New Scientist* (Figure 8.3) makes clear, the mapping sciences are fully embedded within both an economy and a culture of military and security practices. They are fundamentally and deeply surveillant and monitoring sciences, with strong footings in the various training and research centres of the military-information-gathering agencies. In this sense, they are part of the broader political economy of international relations, firmly embedded in and with direct effects on the geographies of modernity. What patterns of differential access and use will emerge within the cyber-worlds of remotely sensed images, global positioning systems and computer-aided mapping? How will system–lifeworld relations be mediated under regimes of electronic accounting, management and mapping? How will personal lives and individual privacy be enhanced or hindered by information mapping technologies? How will the new telematics and informatics contribute to the projects of state monitoring, capital penetration and military enclosure? How will class politics and a politics of difference fare under electronic administration and new representational practices? What new capacities are emerging to sustain local, anti-hegemonic discourses and practices? What forms of cultural politics are enabled in and through these emerging new capacities? And, how are these 'local' expressions of new cultural politics to be thought in the context of globalization, standardization and rationalization of consumer society, mass culture and alienated politics? Mapping is part of the expansion of a particular kind of capitalism that has enabled industrialized countries to make what Samir Amin (1994: 85) has called 'a cheap conquest of the world'. Thus, we need to ask: what are the social implications of digital mapping and the representational regime of techno-capitalism?

TECHNO-CAPITALISM AND DELOCALIZATION[9]

> The 'new utopia' presupposes the disappearance of the local in favor of the spatial. Delocalization means the insertion of the logic of the new communication technologies within the universal history understood as rationalization ... The technical utopia of a society decentralized by telecommunications signifies a spatialization of communication so that all localization becomes impossible. It means the final dissolution of all ties and places that symbolically structured traditional society.
>
> (Raulet, 'The new utopia: Communication technologies')

In earlier chapters we have seen how cartographic practices and forms of social life were complexly articulated. In particular we have seen how cartographic reason, economy and state have been intricately intertwined historically in the production of abstract spaces (of commodities, of private property, of state administration, of judicial power, among others). Underpinning contemporary claims for the democratizing potentials of informatics and new imaging and mapping systems are several key assumptions. One is the western trope of a public space in which people (still apparently predominantly 'men') of good faith join in debate about their future. This promise and possibility of informed open discussion currently serves as a central trope and wish image of the informational economy. It suggests a putative openness of new electronic information media, a rhetoric of 'voice', 'access' and 'information', a trope of reasoned, open, uncoerced discourse in a public place. But here too the mythic image of a democratic culture of debate and negotiation is predicated on individual autonomy, private property and state power. Public space and

the democratic potential of the news and communication media are con-
joined in information systems as the embodiment of civic life. But it is
uncomfortably linked to a project of partial interests and private profit.

Thus, the enabling of the subject, the pluralization of culture, and the
extension of democratic practice these information and imaging systems
portend must also be seen against a backcloth of increasing monopoly
control over information systems and other electronic technologies. Like
all highways, the information highway requires capital investment, points of
access, navigation skills, and spatial and cultural proximity for effective use.
Like the automobile highway, the information highway fosters new rounds
of creative destruction and differentiates among users and between users
and non-users. It brings regions of difference under a common logic and
technology, and through differential access and use exacerbates old and
creates new patterns of social and economic differentiation. While for
some, information means the provision of alternatives and the satisfaction
of choice (even if a 'choice' signifies a socially constructed yet now naturalized whim of the wealthy consumer), for others this post-industrialism (and
its attendant postmodern cultural forms) must still be seen in the context of
a political economy of graft, monopolism and uneven development.

Such processes of territorial colonization, globalization and production
of new scales of action contrast sharply with a techno-cultural ideology of
enhanced autonomy and self-actualization, and severely complicate the
assessment of the relationship between technological innovation and social
change. Not only do data technologies increasingly treat all data and
information within a universal calculus and binary logic, and imaging and
mapping practices reach without break across socially and historically differentiated territories, but the tools themselves permit types of surveillance and intervention that can restructure everyday life itself (see Curry
1998; Goss 1995; Lyon 1994; Pickles 1991, 1995). Mapping and geographical imaging have been fundamentally transformed in the process, providing new exciting and useful ways of mapping worlds.

For some, these new data-handling and mapping capabilities are fully
naturalized as the next logical, efficient, useful and hence necessary, step
in the advance of science and society. They are a stimulus to new ways in
which individuals and groups can overcome the barriers of distance and
enhance their abilities to exercise control over society, space and the
earth. Digital information and mapping systems enhance our understanding, increase the efficiency with which we handle complex spatial data, and
enable new forms of complex relational analysis. In this view, new
remotely sensed imaging and mapping technologies greatly enhance our
ability to understand the earth and society, and they extend our ability to
live democratically and rationally.

Others are more sanguine about the rationalizing effects of such modernizing mapping technologies. For these people, the new systems of
knowledge engineering raise many questions about freedom, civil society

and democratic practice. In part these questions arise because information and mapping systems have become important and independent arenas of commodity production in their own right and (along with hypertext, multimedia and virtual reality) very profitable frontiers for investment. For David Harvey (1989) electronic information systems emerged as a result of changes in the structure of capitalism and the liberal state, as each struggled to deal with fiscal and legitimacy crises of the 1970s. From this perspective, any engagement with questions about the democratic potential of spatial data-handling and imaging technologies will first require a political economy of information and technology in which we better understand how each operates within the broader restructuring of late capitalism. This restructuring is not technologically determined nor driven by markets alone, but it is part of a broader class struggle to create new relations of domination in the workplace, to place in production more efficient technical and organizational practices, to extend patterns of commodification and commercialization into new forms and new niches (including information and data), and to orchestrate new modes of social control and new methods of conducting war (Clarke 1988). In this context of market ideology and deregulated capital, everything comes under the sway of 'information' as an object of counting, measuring and analysis. Modern technological society even sets up human beings and nature as objects of manipulation in such ways that '[o]ur whole human existence everywhere sees itself challenged – now playfully and now urgently, now breathlessly and now ponderously – to devote itself to the planning and calculating of everything' (Heidegger 1969: 34–5).

One key site for the production of these abstract techno-spaces is to be found in geographical information systems (GIS). GIS are particularly powerful and useful computer-based data-handling, analysis and mapping systems that have the capacity for integrating spatial data of any kind: remotely sensed data from satellites and aircraft; areal or topological information about spatial patterns; and discrete data sets that have spatial referents (such as Census data, township, country and state-level data, or site-specific or feature-specific data – point source polluters, production plants, rivers or air currents). Such spatially coded data can be mapped to a common metric and can therefore be overlain with other data sources, and the two can be correlated. Since spatial data may also be mapped at different time periods, GIS also have the capability to produce maps of temporal change across space. Moreover, since the geographical surface to which spatial data is referenced is not itself a planar two-dimensional surface, but exhibits characteristics of altitude, GIS systems have the capability to generate three-dimensional maps of two-dimensional data sets, which, with advanced imaging techniques, permits visual 'fly-overs' of the data surfaces or the generation of complex three- (or multi-) dimensional images.

The 'fly-over' surfaces were, not surprisingly, among the first abstract spaces produced by GIS and put into active use. They provided the capability for generating artificial flight simulators for pilot training and, through further enhancement with artificial-intelligence technologies, also permitted self-guiding drone planes and missiles to be developed based on interactive reading of terrain against pre-set flight plans. As we have already seen, military investment in the development and use of GIS is particularly well developed. Whether for weapons development, general surveillance or basic mapping exercises, GIS have been from the beginning a fully militarized activity, and in part this success stems from the abstract capacities and control systems generated by such data-handling and mapping capabilities.

A different set of abstract spaces has emerged in marketing. Here, mapping digital information permits users to increase the efficiency of operations by, for example, cutting down the costs of printing and mailing of materials to potential consumers and by increasing the effectiveness of each piece of material mailed. Using geographical information systems marketers relatively easily combine discrete data sets (socio-demographic characteristics, police records, credit card and credit-rating records, mailing lists from particular sources), supplement them with specially designed survey information, and produce socio-economic profiles, profiles of recent shopping behaviour, and/or measures of political or social outlook and engagement. This technique, geo-demographic information modelling (GDIM), is already one of the specific techniques in use in targeted marketing campaigns (see Goss 1995; Curry 1998). What is particularly significant about GDIM, however, is the way in which information is collected, aggregated and used. Information on individual purchasing and spending habits is obtained from check-out records, mail-order businesses and credit-rating companies (among others). Since much of this information is geo-coded to either telephone number or post code/zip code, the information can be filed spatially by post code/zip code or neighbourhood. With supplemental material included, GDIM is now able to construct aggregate consumer profiles for specific neighbourhoods. Individual privacy issues and the ethics of compiling locationally specific data profiles are currently hotly debated (see Curry 1998). But there is a further issue of concern. The constructed profiles become the basis for targeted marketing campaigns in which information is differentially circulated to neighbourhoods. Spatially averaged consumer profiles now become the basis for constructing consumers and neighbourhoods begin to be shaped by the targeted commercial and political information they receive based on such profiling.

As the capacities and applications of spatial data and mapping systems continue to be broadened and deepened by forces of cybernetic capitalism and the celebration of technoscience, these questions remain pressing and open. We are only at the beginning of understanding the consequences of these processes of delimiting, mapping, and reterritorializing society and nature with these new abstract spaces. Thus, we need to ask again about the ways in which electronic information and mapping technologies are reconfiguring the contemporary world. As counting machines and type-writers had done earlier, new computerized information systems and artifi-cial neural networks facilitate data entry, capture and reproduction.[10] Informatics effect new capacities in speed, efficiency, and the reduction of effort by which we communicate and act (see Virilio 1986; Virilio and Lotringer 1983). These new forms of experience correspond in part to the shift from a modernist Fordism to a liberal productivism and post-Fordism (Lipietz 1992). They also emerge at the boundary of the cold war, and here information-handling and imaging systems function to create new codes whose liminal futures and new geographies are only now being written. Mapping techniques extend a rationalistic logic – a universal cal-culus – to unify space as object and earth as exploitable resource, unified community or commercial logo. A naturalized present is scripted and inscribed within the domains of cultural production, in terms of which new cultural imaginaries of natural (earth, nature, globe) and social identity are being forged, and electronic images of the earth, interactions in cyber-space, conversations on the community net, or concrete engagements with virtual reality represent self and others in new ways, create alternative forms of experience, and establish new forms of social interaction. Fully normalized, the technics of data exchange and representation legitimize new social practices and institutions, and disseminate a new political economy of the social body in ways that we have only begun to recognize and regulate.

COMMODITY, PHANTASMAGORIA, SPATIAL IMAGE

> A depiction is never just an illustration. It is the material representation, the apparently stabilized product of a process of work. And it is the site for the construction and depiction of social differences. To understand a visu-alization is thus to inquire into its provenance and into the social work that it does. It is to note its principles of exclusion and inclusion, to detect the roles that it makes available, to understand the way in which they are dis-tributed, and to decode the hierarchies and differences that it naturalizes. And it is also to analyze the ways in which authorship is constructed or concealed and the sense of audience is realized.
>
> (Fyfe and Law, *Picturing Power: Visual Depiction and Social Relations*)

In *The Dialectics of Seeing* Susan Buck-Morss (1989: 82) describes Walter Benjamin's use of the term 'phantasmagoria' to describe the spec-tacle of the bourgeois city – the magic lantern show of optical illusions in which consumers were constructed through the display of commodities; a representational economy in which '[e]verything desirable, from sex to social status, could be transformed into commodities as fetishes-on-display that held the crowd enthralled even when personal possession was far beyond their reach. Indeed, an unattainably high price tag only enhanced a commodity's symbolic value.'

The experience of the arcade in which one moved from display window to display window was symbolically reproduced in the panorama – a favourite of arcade paraders – through which they could parade at increasing speed across a wide range of phantasmagoria specially designed for the spectators' edification. For Benjamin (1989: 79), the panorama represented the phantasmagoria of progress: arcades, world exhibitions and the city. As temples of commodity fetishism, each promised in various ways, the possibility of social progress without revolution, each opened new public spaces that sought to mask class antagonisms, and each offered a 'strategic beautification' – the glitter of modernity as immediate proof of progress and the possibility of acquisition. Through these phantasmagoria, monotony is nourished by the new and revolutions are but temporary interruptions which leave the class position of the bourgeoisie unassailed.

Benjamin's reflections on the panorama and the city can, I think, provide us with a productive *entrée* into a consideration of new cartographies. Through this lens, we can investigate how the deployment of an economy of vision through the technologies of commodities-on-display (the store window, the arcade, the panorama, the city street) has parallels in the development and deployment of virtual imaging, mapping and geographical information systems. For Benjamin (1968: 233–4), there were important differences between the image produced by the camera and that produced by the painter. 'The painter maintains in his work a natural distance from reality, the cameraman penetrates deeply into its web. There is a tremendous difference between the pictures they obtain. That of the painter is a total one, that of the cameraman consists of multiple fragments which are assembled under a new law.'

When Benjamin (1968: 223) asks: 'What is the social basis for the contemporary decay of the aura of the image-object?', he answers:

> It rests on two circumstances, both of which are related to the increasing significance of the masses in contemporary life. Namely, the desire of contemporary masses to bring things 'closer' spatially and humanly, which is just as ardent as their bent toward overcoming the uniqueness of every reality by accepting its reproduction. Every day the urge grows stronger to get hold of an object at very close range by way of its likeness, its reproduction. Unmistakably, reproduction as offered by picture magazines and newsreels differs from the image seen by the unarmed eye. Uniqueness and permanence are as closely linked in the latter as are transitoriness and reproducibility in the former. To pry an object from its shell, to destroy its aura, is the mark of a perception whose 'sense of the universal equality of things' has increased to such

a degree that it extracts it even from a unique object by means of reproduction. Thus is manifested in the field of perception what in the theoretical sphere is noticeable in the increasing importance of statistics. The adjustment of reality to the masses and the masses to reality is a process of unlimited scope, as much for thinking as for perception.

I am particularly intrigued by the way in which he presents these differences and what he calls a new law of assembling images. With the emergence of geo-referenced digital data, computer graphic representation and virtual reality, surely the 'law of assembly' has changed again and new forms of perception are being developed. The principle of intertextuality common to both hypertext and information systems directs our attention to the multiple fragments, multiple views and layers that are assembled under the new laws of ordering and reordering made possible by the microprocessor. A strange epistemological 'binary' is born. On the one side is a representational epistemology in which the image is a reflection of nature. On the other side is a manipulative desire which bends nature to the will of humans. In this sense, as well as legitimizing claims to verisimilitude, digital mapping signals the end of mapping as evidence for anything, or at least the emergence of a representational economy whose illusions – Baudrillard tells us – will be so powerful that it won't be possible to tell what is real and what is not.

At stake in these new visualities and new mappings is a reconfiguration of our relation to space. If, as Benedict Anderson (1991: 6) has suggested, 'Communities are to be distinguished, not by their falsity/genuineness, but by the style in which they are imagined' we need to inquire into the ways in which these spaces are being reimagined and how 'we encounter the objects, images and ideas around us' (Rosenthal 1992: 107).

For Walter Benjamin (1968: 221) the coming of the age of mechanical reproduction resulted in the withering of the aura of the work of art in which the fixed-object-image is destabilized in favour of a playful and wilful reproducibility and manipulability, and 'the technique of reproduction detaches the reproduced object from the domain of tradition. By making many reproductions it substitutes a plurality of copies for a unique existence.' Such new practices create new subjects and new capacities for shaping collectives.

In this regard one might argue that the new abstract spaces produced in such ontologies of transparency and manipulability have been so decoupled from their living counterparts that they become something else entirely non-transparent. Perhaps better than 'transparent' (with its implications of a god-trick and an old semiotic transparency to which we have grown accustomed) we need to return to a notion of such images as metaphorical and ideological in that they so thoroughly abstracted from the materiality of lived experience (and in turn redefine both materiality and lived experience).

DIGITAL TRANSFORMATIONS: REIMAGING AND REIMAGINING NATURE AND SOCIETY

In *We Have Never Been Modern*, Bruno Latour (1993) discusses the debate between Boyle and Hobbes in the mid-seventeenth century. Through it he shows how a modern notion of representation came into being with the distinction between science and politics. The Boyle–Hobbes debate stands, in this discussion, for an originary moment from which spring two related but separate notions of representation. Both are underpinned by a single modern anxiety about the need to control the masses. One notion of representation is that which involves the political representation of the views of citizen in an emerging democracy – representative democracy. In this notion of representation, a modern notion of 'Society' is born as that structure of social relations that *must be* represented and regulated politically. The Leviathan will require maps of its territory and information about its citizens and places. A second notion of representation is that which involves the representation of natural objects and in this move 'Nature' as we now know it is produced as a mapped and graphed domain of abstract relations (see the discussion of von Humboldt in Chapter 6). Thus, even our most basic categories of 'Society' and 'Nature' have been produced historically. The 'constitution' of modernity (its historical production and its governing laws) is the structure of science and politics that keeps society and nature distinct and subject to regimes of representation by experts. But, as the title of the book indicates, Latour believes that this constitution and its binary geometry have neither been achieved nor can they be. Instead, the constitution that keeps society and nature, politics and science, representer and represented separate in a kind of symbolic second space has given birth to uncontrollable and unrepresentable monsters and hybrids.

What kind of transition is at work then in this new spatial turn? It is certainly one that puts into question many of the assumptions about mapping as representational, in the sense of a mirror of nature. Instead, we need ways of thinking about geography and mapping that do not presuppose the master narratives of modern cartography and do not hide the politics in science (or the interests behind the map, as Brian Harley, Denis Wood, Robert Rundstrom and Matthew Edney among others have taught us). The task is one of constructing a post-representational cartography.

This is also the possibility for developing an epistemology that Isabelle Stengers has called 'guerrilla' epistemology. For Stengers (1997: 118):[11]

> the problem of the contemporary sciences is not ... one of scientific rationality but of a very particular form of mobilization: it is a matter of succeeding in aligning interests, in disciplining them without

destroying them. The goal is not an army of soldiers all marching in step in the same direction; there has to be an initiative, a sense of opportunity that belongs rather to the guerrilla.

It is the possibility for a renewal of direct democratic practices that destabilize, and have the tools to always challenge, any and all hegemonies – be they created by representational science in the name of nature or by representational politics in the name of society. '[I]t leaves us free to work at modifying these institutions without burdening ourselves with atemporal problems like those of Reason, Understanding, or the West' (Stengler 1997: 118). It opens the possibility for a different epistemology and politics of digital transformations and mappings.

Gillian Rose (1993) has suggested that the conception of the mirror and the imperial eye, so prevalent in the history of modern cartography, is also thoroughly masculinist in nature. I hope by this point in the book that, through the layering of images and arguments, this point has been amply demonstrated. In place of this totalizing and masculinist vision we need to think in terms of different epistemologies of mapping, ones in which – as Rose suggests – the mirror has been broken into a thousand pieces, each shard still reflecting, but without overall coherence, without the possibility of the universal view, without the possibility of control. Is this a future that is possible or even desirable in the 'digital transition/transformation'? Is this a future way of thinking about mapping practice? Is this a new cartography?

George Landow (1992) has come to a similar conclusion in his work on hypertext. For Landow, digital information systems and specifically hypertext promise new ways of theorizing information and representation. The apparently infinite malleability and reproducibility of spatial information in digital systems allows, even forces, us to rethink the relations among objects and practices that have been set in concrete for hundreds of years under the regime of print capitalism (Anderson 1991). Textuality, narrative, margins, intertextuality and the roles and functions of readers and writers are all reconfigured in the digital text. The digital transformations of geo-mapping in Roland Barthes's (1987) terms point to the possibility of the production of writerly (rather than readerly) texts, which do not dominate the reader and insist on particular readings, but engage the reader as an 'author' and insist upon the openness and intertextuality of the text – that is, its openness to other texts and readings. In this way digitality opens up again the question of participation and provides new opportunities for interactivity lost to an earlier nineteenth-century information revolution.

It is to this issue that we now turn. In so doing, I want to begin to read digital mapping, especially geographic information systems and remote sensing as new forms of line drawing; new cartographies for new worlds. That is, I want to ask how these new forms of mapping presuppose and foster new ontologies and practices of transparency and malleable depth; digital information in bits and bytes, 1s and 0s, arranged and rearranged to construct mapped layer upon layer, thematic abstraction on abstraction, enabling filial vectors of association and relation to be mapped one on another; the world rendered as layers, curtains, constellations and flows of potentially infinite manipulability (Figure 8.5).

Martin Dodge has shown us one possibility for such a flexible, malleable and open cartography in his *Atlas of Cyberspaces*. This is a web-based atlas of the linkages, networks and flows of the new geographies created by the information society, what he calls the geographies of the worlds in the wires.[12] Here, nodes, links and networks are rendered cartographically in exciting and innovative ways, as colourful curtains, chains, branches, streams, bits and bytes, connections and interactions, all mapped to render visible the unseen world-in-the-wires. Images such as these represent a fundamental shift in the ontology of objects with the emergence of a new scopic regime of transparency. In this remapping, the penetrating gaze enabled by digital visualization is mobilized not only to picture the world differently, as transparent and clear, but it is also changing the way we understand and use nature and society.

In *Mirror Worlds*, David Gelernter (1992: 1) extends this metaphor of layers, curtains and manipulable flows. *Mirror World* 'describes an event that will happen someday soon. You will look into a computer screen and see reality. Some part of your world – the town you live in, the company you work for, your school system, the city hospital – will hang there in sharp colour image, abstract but recognizable, moving subtly in a thousand

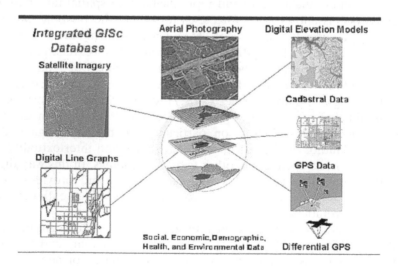

Figure 8.5 Virtual earth: layers and flows

places'. The mirror world of virtual reality and spatial images is a 'true-to-life mirror image trapped inside a computer – where you can see and grasp it whole' (p. 3). These images 'engulf some chunk of reality' (p. 6) and the mirror world 'reflects the real one' (p. 6). 'Fundamentally these programs are intended to help you comprehend the powerful, super-techno-glossy, dangerously complicated and basically indifferent man-made environments that enmesh you, and that control you to the extent that you don't control them' (p. 6).

How is this to happen? How will the 'place' of mirror world permit one to enter, stroll around, and retrieve archival and live-medium information?

> The picture you see on your display represents a real physical layout. In a City Mirror World, you see a city map of some kind. Lots of information is superimposed on the map, using words, numbers, colors, dials – the resulting display is dense with data; you are tracking thousands of different values simultaneously. You can see traffic density on the streets, delays at the airport, the physical condition of the bridges, the status of markets, the condition of the city's finances, the current agenda at city hall and the board of education, crime conditions in the park, air quality, average bulk cauliflower prices and a huge list of others.
>
> This high-level view would represent – if you could achieve it at all – the ultimate and only goal of the *hardware* city model. In the software version, it's merely a starting point. You can dive deeper and explore. Pilot your mouse over to some interesting point and turn the *altitude* knob. Now you are inside a school, courthouse, hospital or City Hall. You see a picture like the one at the top level, but here it's all focussed on this *one* sub-world, so you can find out what's really going on down here. Meet and chat (electronically) with the local inhabitants, or other Mirror World browsers. You'd like to be informed whenever the zoning board turns its attention to Piffel Street? Whenever the school board finalizes a budget? Leave a software agent behind.
>
> (Gelernter 1992: 16–17)

This chimeric world of vision that penetrates solid objects has captured the imagination of scientists dealing with both environmental and social problems. Elsewhere I have called this ontology 'investing objects in depth' (Pickles 1998a). There are many ways in which this new ontology could be shown to be affecting our understanding and use of maps. Indeed, such emergent properties of maps and mapping are quite varied in form as technological and disciplinary developments proliferate opportunities for new imaging and mapping practices. But there is one specific way in which digitality seems to be re-rendering the earth and it is to this form that I now turn.

Gelernter's mirror world captures this new ontology of subjects invested in depth, but does – I think – misname it. Gelernter draws upon the modernist metaphor of the 'mirror' to give a name to this new ontology of subjects in depth. I prefer to place it under the sign of 'visible bodies'; an ontology not of reflective representation, but productive reconstructive surgery in which bodies are mapped in transparent and malleable depth.

This is one of the goals of recent developments in medical imaging. In medicine, imaging systems are producing new representations that render the human body as transparent and in depth,[13] changing the way in which medicine operates. Virtual reality technologies and digital body imaging are enabling a host of relatively non-invasive medical procedures to be developed in treating patients who previously would have required invasive surgery and potentially long and difficult rehabilitation. New legal and institutional arrangements are being established. New visual regimes and epistemologies are emerging that are no longer bound by an ontology of surfaces and bodies. In their places, doctors are beginning to deal with chimera, ghosts and holographs, or with monsters which are reshaping the 'real' (Figure 8.6).

At work in these new imaging technologies may be a distinctly new visual system, a new way of rendering the world-as-picture. Medical imaging and remote medicine are reworking our understanding of the patient's body, and are producing the very mirror worlds of which Gelernter wrote. Here, new objects are doubly invested-in-depth as transparent – 'as utterly available as visible matter' (Waldby 2000: 5).[14]

The notion of a transparent earth has similar enormous rhetorical and visual power. In some ways, Alexander von Humboldt's analytical diagrams in the nineteenth century shared an equivalent goal of rendering the earth abstractly visible. With the imaging and visualization technologies now available, the goal of analytical abstraction and purification can now be accomplished in ways that create abstract spaces of transparent (visible, clear, obvious, unproblematic) objects. Not only are the technical capacities developing quickly for such mapping of the abstract spaces of the earth and society, but so also are the institutions and coordinating mechanisms to make them possible. One of the largest of these new mapping institutions is the US Government's Digital Earth Initiative (Figure 8.7).

The Digital Earth Initiative (DEI)[15] is a coordinated effort on the part of federal government and non-government agencies to construct 'a virtual representation of our planet that enables a person to explore and interact with the vast amounts of natural and cultural information gathered about the Earth' (http://www.digitalearth.gov/main.html). The initiative was introduced in 1998 by then Vice President Gore as a means of mobilizing government and non-governmental agencies to work together to integrate systems for collecting, retrieving and representing

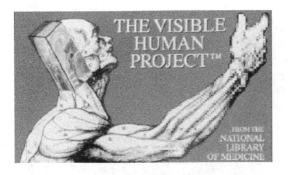

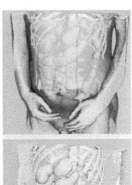

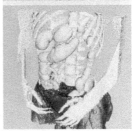

INTEGRA ET   AB OMNIBVS
PARTIBVS   *LIBERA AC*
nuda venæ   *cauæ delineatio.*

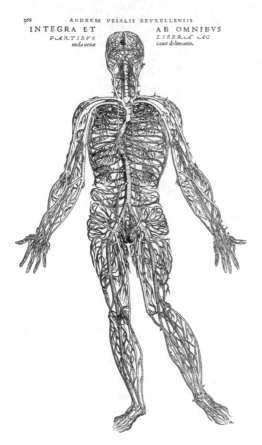

Figure 8.6 'The Visible Human Project™ The National Library of Medicine'
(http://www.nlm.nih.gov/research/visible/visible_human.html) *A Chart
of Veins*, Woodcut attributed to the Workshop of Titian of Vesalius. De
Humani Corporis Fabrica, Basle, 1543

Figure 8.7 'The Digital Earth' (http://www.digitalearth.gov/)

geo-referenced data. DEI is underpinned, in part, by a complex institutional architecture emerging under the National Spatial Data Infrastructure (NSDI). NSDI refers to the integration and sharing of data infrastructure among federal and state agencies. In conjunction with the Federal Government Data Committee (FGDC – http://www.fgdc.gov), NSDI is establishing metadata standards, framework data and a geospatial data clearinghouse for data consistency and data sharing. The framework data provide common standards for data gathering, handling and sharing, and is integrated with the National Geospatial Data Clearinghouse (http://nsdi.usgs.gov). DEI is specifically charged with addressing technical and organizational difficulties with managing and making available high-quality geo-referenced data through programmes on computational science, mass storage, satellite imagery, broadband networks, interoperability and metadata. Gore outlined the goal in terms very reminiscent of *Mirror Worlds* in the following way:

Imagine, for example, a young child going to a Digital Earth exhibit at a local museum. After donning a head-mounted display, she sees the Earth as it appears from space. Using a data glove, she zooms in, using higher and higher levels of resolution, to see continents, then regions, countries, cities, and finally individual houses, trees, and other natural and man-made objects. Having found an area of the planet she is interested in exploring, she takes the equivalent of a 'magic carpet ride' through a 3-D visualization of the terrain. Of course, terrain is only one of the numerous kinds of data with which she can interact. Using the system's voice recognition capabilities, she is able to request information on land cover, distribution of plan and animal species, real-time weather, roads, political boundaries, and population. She can also visualize the environmental information that she and other students all over the world have collected as part of the GLOBE project. This information can be seamlessly fused with the digital map or terrain data. She can get more information on many of the objects she sees by using her data glove to click on a hyperlink. To prepare her family's vacation to Yellowstone National Park, for example, she plans the perfect hike to the geysers, bison, and bighorn sheep that she has just read about. In fact, she can follow the trail visually from start to finish before she ever leaves the museum in her hometown.

She is not limited to moving through space, but can also travel through time. After taking a virtual field-trip to Paris to visit the Louvre, she moves back in time to learn about French history, perusing digitized maps overlaid on the surface of the Digital Earth, newsreel footage, oral history, newspapers and other primary sources. She sends some of this information to her personal e-mail address to study later. The time-line, which stretches off in the distance, can be set for days, years, centuries, or even geological epochs, for those occasions when she wants to learn more about dinosaurs.

(Gore: http://www.digitalearth.gov/VP19980131.html)[16]

In this transparent ontology of mapping, the earth and its spaces are thoroughly plastic and manipulatable. They can be viewed from any angle, in any available spectrum, with whatever categorical or technical filters one needs for any particular purpose. In this sense, mapped earthly objects permit what phenomenologists have called 'adumbration' – the multiple elaborations of the objectness of a particular object from a variety of different positions or perspectives. In part, I think this was the strange fascination phenomenology held for Peter Gould as he continued to push his spatial analytics beyond the representational logics of what he variously called the Pythagoreans, Linneans and paleontologists (Figure 8.8). For Gould, the representational and categorical logics of these mappings, with all their assumptions about linearity and correspondence, were to be cut loose by the turn to mathematical languages of mappings (surjective,

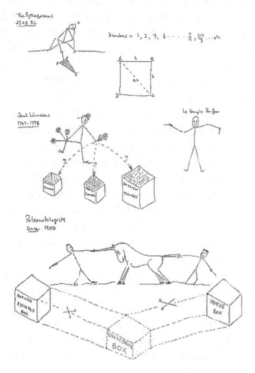

Figure 8.8 Peter Gould's mappings (with permission, Jo Gould)

injective, bijective and relational). The resultant multidimensionality of the spaces of relations, mappings, and functions opened a new window on the earth and society as objects of our care (*Sorge*) and attention. It was only through abandoning the metaphysics of presence and the ontotheologies of realism and representationalism that truly human geographies could emerge. That there was resistance to the fact that such geographies might be mathematical and highly abstract was, for Peter, only a sign of timidity on the part of geographers and reluctance on the part of those unwilling to 'release' into the deeply human world of qualitative mathematics and multidimensional spaces. Peter was fascinated by any innovative mapping practice that respaced the world, be it mental mapping, the exploratory cartography of Waldo Tobler, the abstract materialism of Gunnar Olsson, the reconstructed spaces of archaeology, or the spaces of capital. He would, I think, have been equally excited about the spatial reconfigurations occurring in the Digital Earth Initiative and also in the Vegetation Canopy Lidar (VCL) project (Figure 8.9).

   For many years, scientists using either ground-based or remotely sensed research methods have been puzzled by how to measure total biomass in complex forest environments. In such environments, access to the com-

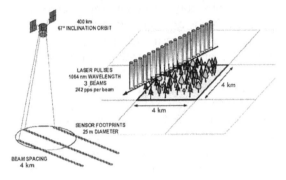

Figure 8.9 Vegetation Canopy Lidar Mission (NASA. http://essp.gsfc.nasa. gov/vcl/)

plexity of their vertical structures has been extremely difficult and has relied on painstaking mapping based on sampling from underneath (or occasionally in) the canopy and reflectance measures of the canopy surface. However, informed policy judgements about global warming, for example, require detailed knowledge about the processes of forest growth and biomass exchanges. For such measures, traditional ground-based and remote surveillance methods have proven inadequate.

The Vegetation Canopy Lidar (VCL) project is run by Ralph Dubayah and colleagues at the University of Maryland and is funded by NASA's Earth System Science Pathfinder (ESSP) programme (http://www.geog. umd.edu/vcl/). Its goal is to develop a new system of earth imaging that will produce the first three-dimensional mapping of the vertical structure of the earth's vegetation cover and land surface. The project uses

> 5 diode-pumped, Q-switched Nd:YAG lasers, generating 15 mJ, 10 ns-wide Gaussian shaped pulses at a wavelength of 1064 nm. The lasers operate at frequencies of 10 Hz (over oceans) and 242 Hz (over land). Lasers are in a circular configuration which from a 400 km-high orbit will span an 8 km-wide area. Laser footprints are 25 m wide. They are near contiguous in the along track direction, and spaced 2 km apart

across track ... Global Positioning System (GPS) and Satellite Laser Ranging (SLR) techniques provide the spacecraft orbit to 15 cm (1 sigma) accuracy.

(http://www.geog.umd.edu/vcl/)

The goal of VCL is to render the forest visible in depth and – as with the Visible Human Project – to open new abstract spaces of information. This is a project to produce a new second nature – an abstract mathematical modelling and mapping that gives the impression of a transparent holographic earth.

As the hard work of imaging the earth in this way continues, a broad cultural economy of vision in depth may be being constructed. In holography, Virtual Reality (VR) machines and full-immersion environments, this impulse reaches its zenith, both in terms of the technical advances being made and in the commercial and military interest in them (see Hillis 1999a, 1999b). Full-immersion environments and VR simulations push these ontologies of transparency to their full extension and, like the examples we have discussed, invest their subjects in depth. Gone are ontologies of non-penetrable surfaces and objects. Now all objects are bundles of information that can be imaged in as many ways as they can be imagined: the boundary layer of clouds is wiped 'clean' from 'cloud-free globes' (http://www.earth-images.com/haz.htm), the surface boundary of the earth is rendered plastic and transparent, and the forest canopy is rendered virtually accessible.

DEMOCRATIZING THE NEW CARTOGRAPHIES?

> The burning question, then, becomes this: why have the immense processual potentials brought forth by the revolutions in information processing, telematics, robotics, office automation, biotechnology and so on up to now led only to a monstrous reinforcement of earlier systems of alienation, an oppressive mass-media culture and an infantilizing politics of consensus? What would make it possible for them finally to usher in a postmodern era, to disconnect themselves from segregative capitalist values and to give free rein to the first stirrings, visible today, of a revolution in intelligence, sensitivity and creativity?
>
> (Guattari *Incorporations*)

The 'democratization' of the image and information, and the corresponding adjustment of the masses to this new reality, have important implications for how we understand space, society and nature. We have seen some of the ways in which digital mappings have created more extensive mechanisms of surveillance, enabling greater powers of reach and control over territory, society and nature (Figures 8.10 and 8.11). In what ways have they

**EVERYTHING THAT'S OUT THERE
CAN BE FOUND RIGHT HERE.**

Figure 8.10 Digital Earth: 'Everything that's out there can be found right here'

also created new potentialities for social action and new configurations in social life? What are the effects of these changes, especially in circumstances in which the technologies of reproducibility can – in the hands of a Brecht or a Dadaist – be turned into critical theatre or art which challenges and destabilizes the categories and arrogance of bourgeois culture and life?

Central to all these cybernetic systems and their corresponding cyberspaces are the emerging geographical and geopolitical spaces of late capitalism.[17] The geopolitics of cyberspace is first and foremost an imperial geography of the resource-rich West/North/First World, deployed in the specific interests of economic revitalization and capital accumulation.[18] As Donna Haraway (1989) has suggested, this imperium of the West/North/First World is 'the one who is not animal, barbarian, or woman; man, that is, the author of a cosmos called [H]istory'. But it is also, as we have seen, a possible cartography of anti-imperial spaces that generate hope for 'a family of figures who would populate our imagination of these postcolonial, postmodern worlds that would not be quite as imperializing in terms of a single figuration of identity' (Penley and Ross 1991: 18). I turn now to

Figure 8.11 'The art of prospecting for customers' (with permission, GeoTech Media www.geoplace.com)

a consideration of these cartographies of anti-imperial spaces of hope and to their multiple figurations of identity.

In considering the effects of earlier technologies of representation Walter Benjamin (1968: 241), suggested that the political economy of capitalism and the emergence of technologies of mass communication presaged not only an opening of the public sphere, but also a politics and aesthetics of the masses that supported war, and for very particular reasons:

> Mass reproduction is aided especially by the reproduction of masses. In big parades and monster rallies, in sports events, and in war, all of which nowadays are captured by camera and sound recording, the masses are brought face to face with themselves. This process, whose

significance need not be stressed, is intimately connected with the development of the techniques of reproduction and photography. Mass movements are usually discerned more clearly by a camera than by the naked eye. A bird's-eye view best captures gatherings of hundreds of thousands. And even though such a view may be as accessible to the human eye as it is to the camera, the image received by the eye cannot be enlarged the way a negative is enlarged. This means that mass movements, including war, constitute a form of human behavior which particularly favors mechanical equipment.

Benjamin's discussion of the relationship between technologies of abstraction, expression and representation and the politics of the day have eerie parallels with the emergence of new technologies and the frequency of war in the late twentieth century. What we do not yet understand very well, however, are the kinds of politics and aesthetics that are emerging and the social forms to which they correspond. If mass communications and the camera enabled the production of the masses in a particular way in the mid-twentieth century, what kinds of social subjects and what forms of the multitude are being produced today with our own systems of representation? The incorporation of new mapping technologies in the Gulf War 'pin-point' blanket-bombing of Iraq, their uses in the reterritorializing of ethnically defined states at Dayton, neighbourhood-marketing strategies, political gerrymandering, environmental activism, community development, and their presentation through television as ritual celebration of national power and virtue, all raise important questions about the processes of subjectification and the production of spaces to which they contribute.

As Benjamin had asked of photography, Allucquere Rosanne Stone (1995) asks what is happening in the deployment of emergent digital technologies, what new forms of identity are being produced, and what kinds of 'counter-images' are available to us?

> *The War of Desire and Technology* is about science fiction, in the sense that it is about the emergent technologies, shifting boundaries between the living and the nonliving, optional embodiments ... in other words, about the everyday world as cyborg habitat. But it is only partly about cyberspace. It is also about social systems that arise in the phantasmatic spaces enabled and constituted through communication technologies ... I am interested in prosthetic communication for what it shows of the 'real' world that might otherwise go unnoticed. And I am interested because of the potential of cyberspace for emergent behavior, for new social forms that arise in a circumstance in which *body, meet, place*, and even *space* mean something quite different from our accustomed understanding. I want to see how tenacious these new social forms are in the face of adversity, and what we can learn from them about social problems outside the worlds of the nets.

The dislocation of universals and of the authority of the text is at the same time an opening up of the intertextuality of all texts. Digital information and imaging systems multiply embedded systems of texts (including signs, databases and representations).[19] All texts are, in this sense, embedded within chains of signification: meaning is dialogic, polyphonic and multivocal – open to, and demanding of us, a process of ceaseless contextualization and recontextualization. Intertextuality, in this sense, cannot be fused with positivist or more broadly empiricist epistemologies, but requires a thoroughly different understanding of epistemology – a rejection of the univocity of texts (and images), of representation as a mirror of nature, and of a metaphysics of presence (and the foundational claims of positivism) to ground itself unproblematically in the given real world or the immediacy of observation. The implications of such a radical contingency and thoroughgoing contextual and 'disseminated' notion of representation are devastating for so much of what Fredric Jameson (1971) has called the 'anti-speculative bias' of the liberal tradition with its 'emphasis on the individual fact or item at the expense of the network of relationships in which that item may be embedded'.

Intertextuality implies a decentring of the author and the reader, and the situating of meaning in the margins between texts and writers – in an illimitable chain of signification, a network that as Heinz Pagels (1989) suggests has 'no "top" or "bottom" ... no central executive authority that oversees the system'. In this sense, the turn from the text as isolated object to a text embedded in a constantly expanding chain of signification means for information systems, an escape from an idolatry of the work conceived of as a closed, complete, absolute object (Genette 1982). Implied in this shift to an understanding of texts and images as constantly open to interpretation and critique is a different understanding of the work they do. This is a shift from 'readerly texts' (whose purpose is to create readers for already written texts) to 'writerly texts' (whose purpose is to see texts as producing an open series of readings, each of which requires that the reader also be in part author of meaning). Can we think of 'writerly maps' in the same way?

The shift that occurs from the 'tactile' (pen and ink) to the 'digital' (electronic code) combines fixity and flexibility in new ways (Landow 1992: 19). In this sense, Baudrillard (1983: 103) argues, '[d]igitality is with us. It is that which haunts all the messages, all the signs of our societies. But this digitality also brings with it a commitment to a binary logic connected to a particular metaphysical principle: cybernetic control ... the new *operational* configuration.' Digitality and representational technologies produce new codings and practices, and with them new individual identities, social relations and cultural imaginaries. But no matter how real they appear, they are images constructed with cybernetic logics and architectures. These instrumentalities have effects. They too are images and mappings created by the distinction made in lines drawn between 1s and

0s, on or off, present or absent, and the lines of interpolation and extrapolation. The effects of the logics, mathematics and architectures of digital imaging and mapping systems are far too complex and varied to be addressed here. Indeed, they constitute a different history of abstract architectonic spaces that cries out to be written, and one that is rapidly being reworked by the mathematics and technologies of complexity.

In many arenas, technologies of visual representation are challenging our fundamental categories of objectness, clear sight, the seen and the unseen, the obvious, the stable, and the exterior and interior: in the secret labs of the air force where new human–machine interfaces are being operationalized; in the studios of innovative television producers where virtual realities and cyberspace are being 'concretized' and rapidly disseminated as 'real futures'; in ads for sports shoes, soft drinks, or political parties where new cultural codings are occurring; in the operating theatres of hospitals where digital and spectral-imaging techniques are combined to produce real-time internal pictures of organs at work; and in planning offices or the lecture rooms of art departments where students are being challenged to extend the visual powers of cyberart and virtual worlds.

Not only are we entering into the heavens as all-seeing observers, but the sky and earth themselves are opening up for us. Paul Virilio (1997: 3, 6) has echoed these sentiments in *Open Sky*, in which he suggests that:

> Everything is being turned on its head at this *fin de siècle* – not only geopolitical boundaries but those of perspective geometry . . . Appearances generally and those of art in particular are being deconstructed – but so is the sudden transparency of the world's landscape.
>
> Soon we will have to learn to fly, to swim in the ether . . . Without a distant horizon, there is no longer any possibility of glimpsing reality; [we drop into the time of a fall akin to that of the fallen angels and the earth's horizon then becomes just another 'Baie des Anges'.] Philosophical let-down in which the *idea of nature* of the Age of Enlightenment is eradicated, along with the *idea of the real* in the age of the speed of light.

But in what ways are the social, geographical and material circumstances that support this metaphysics of open sky changing the nature of the worlds in which we live?

Part V

Conclusion

The whole practice and philosophy of geography depends upon the development of a conceptual framework for handling the distribution of objects and events in space.

(Harvey, *Explanation in Geography*)

Part V

Conclusion

9 Counter-mappings

Cartographic reason in the age of intelligent machines and smart bombs

To the young practitioners of this new art, the old geographers believe in a flat earth – two-dimensional, static, and Euclidean, akin to a page in a book. The new view is three-dimensional, organic, and Mandelbrotian, akin to a moment of video. The old craftsmen worked with paper, ink, and a list of coordinates; the new breed has massively parallel computers crunching ever-expanding lodes of information. The veterans believe that they have limned a landscape that is knowable; the punks are anxious to discover and map new realms of dynamic ambiguity.

(Hitt, 'Atlas shrugged: The new face of maps')

He inquired about the geological structure in his landscapes, convinced that these abstract relationships expressed, however, in terms of the visible world, should affect the act of painting. The rules of anatomy and design are present in each stroke of his brush just as the rules of the game underlie each stroke of a tennis match. But what motivates the painter's movement can never be simply perspective or geometry or the laws governing color, or, for that matter, particular knowledge. Motivating all the movements from which a picture gradually emerges there can be only one thing: the landscape in its totality and in its absolute fullness, precisely what Cézanne called a 'motif'.

(Merleau-Ponty, *Sense and Non-sense*)

What are the implications of seeing our world as not only produced by maps, but decoded, recoded, and further decoded by many maps and their attendant social interests over time? What does it mean to think geographical and social identities in terms of ongoing processes of socio-spatial decoding, recoding and over-coding? What, after all, does it mean to stand in your garden and watch the squirrels tumble through trees in autumn? Architectural lines, built walls, designed windows, bounded decks, easements and rights of ways, floodplains (insured and not insured), electric wires crossing a near-sky while contrails stretch across the higher skies, and beyond them silent tracks of satellites keeping inner and outer space under surveillance, road systems behind the neighbours' property, fenced along the survey line. Leaving the garden to go to pick up the post/mail, walking up the utility easement – the hidden complex of sewers, drains, and wires tracking in and out of platted and plotted houses, roadways, and streams – to the roadside, the liminal boundary of public and private space (itself deeded, over-deeded, surveyed and oversurveyed, delimited and redelimited, and designated and re-designated on map and document). In this liminal space (perfectly symbolized by the US mailbox, owned and erected by the property owner but controlled, regulated and specified by and for the sole use of the US Mail), letters arrive from banks offering credit cards based on zip-coding databases that track and map purchases and payments. Geo-referenced databases give complete strangers more information about me in two minutes than my friends and families will learn in thirty years. Map after map, layer after layer, identity after identity, combining and recombining, crashing and compounding, erasing and reconfiguring ... sedimentations, striations, inscriptions, projections, gorings, scalings ... markings on the multi-subject that is walking through the garden to check the mail. Codings and recodings producing subject and world along axes of difference, as dwelling, access, flow, consumer, owner, borrower, neighbour; identities and codings that multiply subjectivities in interesting and always unexpected overdetermined ways. We are, in this sense, over-coded as multiply coded shifting, decentred identities. That is, we are rhizomatic:

> The rhizome is altogether different, *a map and not a tracing* ... What distinguished the map from the tracing is that it is entirely oriented toward an experimentation in contact with the real. The map does not reproduce an unconscious closed in on itself; it constructs the unconscious. It fosters connections between fields, the removal of blockages on bodies without organs ... The map is open and connectable in all of its dimensions; it is detachable, susceptible to constant modification. It can be torn, reversed, adapted to any kind of mounting – reworked by an individual, group, or social formation.
>
> (Deleuze and Guattari 1987: 12)

What would it mean, then, if our maps were indeed schizoid, evoking the new realms of dynamic ambiguity and the Cézannean 'motif' with which I began this chapter?

The post-war history of technical development, combined with the longer, western history of observer epistemologies, have produced a highly efficient and widely used science of mapping predicated on technical instrumentation, accuracy and representation aimed at mapping these social, economic, political and geo-strategic relations. On the other hand this scientific cartography has in the process and in important ways written out its own social, cultural and institutional histories and commitments. The complex interweaving of descriptive and perspective scopic regimes has worked to create cartographies with global reach, planetary conscious-ness, and a commitment to unfettered criticism and openness to the world. But this global view and its attendant critical openness, with all its power-ful universalizing and distancing perspectives, has also overlooked or hidden its social commitments and interests, particularly its repressive and productive ties to state, corporate and military power. This combined technical–historical project has been one in which Cartesian–Kantian conceptions have framed science and space as world-as-picture, the God-trick of the all-seeing eye, at one and the same time viewing all places from some particular position of privilege (Metropole, Europe, Male, White, North, Wealthy, Industrial, Urban).

Second, in the hands of post-structuralists the crisis of representation (in part produced by the very reproducibility and manipulability of mapped images, as Hitt indicates in the quotation that begins this chapter) has opened up new sites and questions for a reinvigorated mapping studies – a cultural studies of mapping. The very issues overlooked or hidden by traditional statist and institutionalized cartographies, as a result, have become the subject of intense scholarly interrogation. While cartographers renew their commitments to the business of pursuing the technical 'march of progress', within cultural studies and science studies the origins of mapping techniques in land surveys, the role of imperial projects of terri-torial expansion and control, the ordering and disciplining roles of national topographic mapping agencies, and the rendering of nature and society as objects to be represented graphically as well as scientifically and politically, have all become sub-fields for critical analysis.

Third, we have seen how, in this rethinking of how maps work, some fundamental assumptions about vision and representation have also been brought into question. One result has been that a geographical imagina-tion has begun to destabilize universalist and totalizing visions of mapping and cartography, producing in their stead geographically and historically specific understandings of scopic and representational regimes. The idea that vision and mapping have their own intrinsic geographies is in turn reshaping science and technology studies in new and interesting ways.

Fourth, the reinvigoration of a particular history of representation by a geographical imagination is also tied to the challenge Brian Harley gave us to study maps in human terms, to unmask their hidden agendas, to describe an account for their social embeddedness and the way they func-tion as microphysics of power, and to analyse how they are part of a domain of social practices whose effects have ethical implications for the societies in which we live. As Brian understood so well, when these *broader* social contexts are forgotten, as they have been in much scientific cartography, power is exercised without mediation or reflection and the public sphere is distorted. In 1986, I argued that a society gets the kind of geographical education it deserves; that a democracy that shuns deep geo-graphical engagement and practice has little or no interest in fostering a critical geographical imagination. This was Brian Harley's concern over the generalizing of digital cartography and geographical information systems. We both felt that what appeared to all intents and purposes to be a debate about epistemology, technique and information, was actually central to the kind of society we produce and reproduce, and specifically the kind of democratic possibilities that are forged and protected in the public sphere. It was never the case that geographical and mapping practices were unimportant. Far from it. It was the case that the significance of cartographic and geographic reason in the structuring of modern economies, states and lives was, for whatever reasons, overlooked. As Gunnar Olsson and Frank Farinelli remind us, it was a pity that Immanuel Kant, a teacher of geography for over thirty years and author of the *Critique of Pure Reason*, *Critique of Practical Reason*, and *Critique of Judgment*, never did write the fourth critique; *Critique of Cartographic Reason*.[1] It is perhaps also to our shame that for so long we geographers allowed the cartographic and geographic imaginations to be written out of discussions of social life.

Harley's concern with this issue can, I think, be equated with Juergen Habermas's (1994, 1997) concern with the role of the public sphere at the point of transition and reunification in 1991. Habermas argued that the achievements of the Federal Republic of Germany after 1945 (consolidated and extended after 1968) had led to social gains and specific controls on the predatory nature of capitalism. These needed to be protected after the reunification process of 1991, a fact prefigured in the constitution. Recognizing its own origins in a divided Germany, the Federal constitution required a renegotiation of its own basic principles at the point of reunification. But this constitutional imperative was ignored as the map of a united Germany was redrawn after 1991. Habermas saw this as both a moral failure of the Federal Republic's politicians and a historical mistake that would extend the imperial power of the West over the East, unleash new forms of predatory capitalism, and deepen the processes of uneven development within the new country.

Harley saw a similar ambiguous danger in the emergence and wholesale acceptance of computerized information technology, and especially GIS. Although maps were, he argued, instruments of power and embedded in social systems of ethnocentricism, privilege and control, they were also ambiguous objects as a result of their widespread dissemination by the state and the printing presses. The national topographical paper map and its variants had been disseminated to a broader public, popularizing and democratizing the topographic map (see Matless 1999). Ironically, it is Harley (1990: 1) the theoretician of the power of maps who argued most directly and strongly for retaining topographical maps in their paper form and against the possibility of 'going digital' 'on the grounds that they can offer a democratic and humanistic form of geographical knowledge'. The emergence of digital cartography and GIS required, in his view, a wholesale renegotiation of the relationship – the modern constitution – between technological science and society.

In the ten years since Harley wrote those words, computer mapping has become ubiquitous in societies in which the topographic map was widely available and disseminated. But it has also begun to emerge in important ways in societies without available topographical maps or where topographical maps did not function as democratizing tools. State socialist countries experienced their own crisis of representation in 1989 as popular forces rejected the apparatchik control over information; environmental data, financial accounts and topographic maps were classified as top secret in nearly all state socialist societies, serving few direct roles in the body politic except through agencies of the military and the state. One consequence has been a rapid, albeit ambiguous, adoption of GIS and digital-imaging and mapping systems in recent years leading to a widening and a democratizing of access to information, but also an entrenching of bureaucratic and centralizing tendencies within the planning system (see Pickles and Mikhova (1998) on the role of topographic maps in state socialist Bulgaria and digital mapping in post-socialist Bulgaria. Also see Ben Orlove (2002: 20ff.) for a parallel discussion of the role of maps in Peru). Non-representational mapping cultures encountered their own crisis of representation as one of 'imposition'; textual representations such as topographic maps, with all their attendant objectifications and erasures, were literally and figuratively imposed on their lived worlds creating deep social and economic crises.

Current interest in counter-mapping and local access to global mapping technologies reflects these imperatives to respond to the rapid expansion of information, imaging and mapping technologies. But these responses are also ambiguous as social struggles over natural resources and resource extraction, for example, are increasingly waged with scientific mapping tools such as GIS and remote sensing.

It is to these that I turn in this final chapter by asking the question: how have cartographers (of the paper form and the digital form) begun to think beyond the unmasking of the silences in traditional maps to the production of new maps for new worlds? To what extent have new critical cartographies emerged as a form of deconstructive practice, disseminating, deferring/differing and recontextualizing the world in the interest of a broader democratization of knowledge and information? And, to what extent have these 'Other' crises of representation been considered? In so doing, I want to resist the temptation to read the new cartographies and critical theories as somehow transgressing traditional boundaries, as if those boundaries were themselves univocal and hegemonic. If Enlightenment cartography was always highly contested and conflicted, then the question of counter-mapping must itself be rethought.

CRITICAL CARTOGRAPHIES

In his 1999 book, Marcus Doel asks us to consider the possibility of a cartography that shimmers. Doel seeks to dislodge our commitments to solid and fixed identities, and instead asks us to think about ways in which flows, relations of difference, and change can be mapped. He asks us to begin to think objects as bundles of relations and challenges us to think of a cartography appropriate to such objects. Before turning more fully to this de-ontologized cartography, and before picking up its implications for nomad cartography, I want to first return to Bill Bunge. In so doing I want to suggest that a de-ontologized cartography is not just about new forms of cartography, new representational practices, and the rendering of new objects. It is also about de-ontologizing cartography-as-we-know-it. That is, it is about both the alternatives to Enlightenment cartography (e.g., postmodern cartographies) *and* it is about the *dissemination of cartographies*; a post-representational account of actually existing cartographies. *It is not only that the instrumental logics and representational epistemologies of universalist cartography are to be countered by new mapping forms, but that the discursive practices of modernist cartography are to be deconstructed and read differently.* In so doing, I ask whether it might be the case that the counter-mappings we seek have been with us all along.

In some ways, Bunge prefigures our concern with nomad cartography. But his cartography is nomadic in a very particular way. By all accounts, Bunge was himself a nomad. Cast out from the hallowed halls of academia, Bunge was what Zizek (2001: 1) calls one of the 'free-radicals' neutralized 'to help the social body to maintain its politico-ideological good health'. In his nomadism, Bunge established the Detroit (and later Toronto) Geographical Expedition to bring radical geographers into the inner city to work together with local groups struggling for civil and environmental rights. In encouraging expeditionary geographies that adapted the skills and insights of geographers to socially relevant issues, Bunge suggested that existing mapping practices could easily be adapted to the concerns of the poor and powerless. In this way, the geographical expedition was to be a reclaiming of the traditional geographical claim to expertise, especially to mapping. Geography had for too long worked in the service of the state, empire and capital. Why could the skills so sharply honed at the workface of capitalism and the state not also be used to benefit ordinary people in their everyday struggles against pollution, underinvestment in social and material infrastructure, against physical danger, and against the diktats of urban planning machines? To this end, Bunge insisted on 'the use of any means necessary' to fight for basic human rights in the city and globally. One central tactic of this urban insurgency was to use cartographic methods to fight for particular causes.

What was particularly interesting about this deployment of cartographic skills was Bunge's tactical commitment to using whatever means were at hand to the best of his ability (to deploy the latest mapping and analytical techniques in the struggle), to proliferate their use (the many thousands of maps he produced to demonstrate the alternatives that Detroit Education Board had in making their decisions about school zoning), and to develop and use these techniques in consultation with the local individuals and groups most in need of them.

Bunge was committed to science as a tool of progress or 'critical modernism' (see Pickles (2001, 2002) and Peet and Hartwick (1999, 2002) for a broader discussion of critical modernism), but he was also committed to a pragmatics of map use.[2] He aimed to challenge the traditional fetishes of cartographic and planning practice. Uneven distributions of income, health and education were illustrations of the extreme pathologies of a society and a 'measure of the degree of biological breakdown among the species *Homo sapiens*' (Bunge 1975: 149). In his cartographies of Detroit, abundance and lack, super-abundance and brutal poverty are depicted side by side, whose boundary is 'an intermediate zone in constant danger of falling into poverty' (p. 150). Organic instability, violence, tension, starvation and desperation populate Bunge's cartographies, as he asks us to consider the simple (yet often overlooked) geographical question: how can children go hungry when 'overabundant' food is stored in warehouses, where it often is allowed to rot? How can a cartographic imagination assist those in dire need in Detroit's 'City of Death' to achieve their species-being and the equality that is their right?

Such an insurgent cartography required the adoption of different 'perspectives'. Instead of the rationalizing 'God-trick' of the universal gaze, Bunge insisted on a repositioning of the cartographer vis-à-vis those being mapped. Instead of mapping from the point of view of the urban planner, he insisted on community-based mapping. One result was *Fitzgerald*, a geographical biography of a neighbourhood with a very different social cartography of urban life. Embedded in needs and struggles identified as important to the community, expedition members literally 'mobilized' cartography to make visible the conditions of existence of the ghetto, unemployment, and social conflict. Instead of preparing maps from the planner's point of view or from the 'adult's-eye' view, Bunge (1971, 1975) used abstract mapping to unveil inequality and social violence, committing one project to the preparation of maps relating to children's safety. Simple maps of hazardous materials along streets, incidences of rat-bites, or unlit alleyways would provide useful tools for empowering communities to improve the lives of their children; to literally and figuratively take back the streets. The resulting maps are powerful and poignant images, not the least because they are stark reminders of how few cartographies have – until recently – actually taken a stand in this way, and how much of modern cartography is focused on other objects and interests.

In *Ban the Bomb: The Nuclear War Atlas* (1988), Bunge again asserted the power and necessity of a geographical imagination in dealing with the terror of strategic planning based on mutually assured destruction. In contrast to 'geographers' near criminal neglect': 'Geography's intrinsic insight is that there are plenty of half-lives in physics' infinite time in which to recover from radioactive war, but no place on our finite earthly domain in which to do so.' The atlas is a tight, technical presentation of the geographies of potential impacts of nuclear war. In his 'geography of civilization' he asks, what happens when the upper half of the urban hierarchy is destroyed by nuclear blasts? What are the distance decay curves for the blast, heat and radiation effects of a nuclear bomb, and what are the trend lines of speciescide, nuclear proliferation and weapons accumulation? Deploying the tools of demography, economics, and spatial/urban analysis, Bunge 'enlivens' the geographies of nuclear war. But nowhere is this political dissemination of technique more effective than when he turns to cartographic methods.

> Therefore, gentle reader, read on and then after the hour it takes to study this atlas, act for peace as if the lives of the children in your family, and your own personal life too, depended upon it. To save humanity, save the children from nuclear war!

Today we perhaps reread the history of modern geography too much from the perspective of the end of century, the end of the cold war, and 'the end of history'. But Bunge reminds us of a different time and place when the children of 1968 saw history as a barrier to social progress and the future as open with possibilities (see Watts 2001). If spatial analysis, cognitive–behavioural approaches, and humanistic geographies were all grappling with Cartesian–Kantian problematics, presupposing notions of science, space, subject and mind, that have all proven to be too instrumental, too captured by a cartographic anxiety, they were also struggling with the historical challenges and opportunities of post-war change.

While Bunge's voice was in many ways a voice from the margins, it was also representative of others who were grappling in their own ways with what possibilities there were for more humane, less instrumental, people's geographies. I take this to be precisely the point of Gunnar Olsson's, Peter Gould's, David Harvey's and Derek Gregory's prolonged struggles with spatial analysis. Positivism has always had at its core a fundamental ambiguity: a progressive epistemology and commitment to the democratizing of science, even as it has pushed hard for the instrumentalizing of society and the need to legislate the masses through a cadre of technically trained experts and elected officials. But also, at its very heart, spatial analysis understood the crucial problematics of mapping: that the construction of parametric and non-parametric spaces was an infinitely open analytical exercise, that the world was never narrowly reflected in the mirror of the map, and that the spaces of our lives were limited only by our ability to imagine and draw the lines needed to give them identity. Their fascination with mathematics and theoretical abstraction seemed to offer new flights of imagination to configure new spatialities and new cartographies, as I think their collective fascination with Torsten Hagerstrand's time geography mappings illustrates.

In the United States, the uses of mapping for local empowerment have grown rapidly in recent years. Doug Aberley's (1993) *Boundaries of Home: Mapping for Local Empowerment* is one such example that uses alternative mappings to support bioregionalists' efforts at 'reinhabiting place' (p. 3). Extending this sense of cartographic geosophy and Bunge's expeditionary geographies, Cravey *et al.* have recently returned to the question of the progressive potential of everyday mapping. In turning to what they have recently called the 'mundane experience' of everyday life, Cravey *et al.* (2000: 229) have in essence asked, what would a cartography of experience look like if it turned its attention to at-risk populations? In their essay 'Mapping as a means of farmworker education and empowerment', they develop the ideas of the Brazilian-born scholar–activist Paolo Freire, who sought to change how popular education treated everyday life. Drawing on the experiences of peasants and workers, Freire developed literacy programmes that helped people to increase their control over their personal and community lives, literally by giving them command over their language. Cravey *et al.* suggest that mapping too can operate as a kind of graphical and spatial *conscientization*. Mapping can, in effect, be transformative in both diagnosing and dealing with health issues among at-risk and underserved populations.

In *'Terrae incognitae'* J.K. Wright (1942: 83) was concerned with the closing of geographical categories wrought by totalitarianism ('Map Makers are Human') and by the emergence of a parallel instrumentalism in the social and geographical sciences. Wright urged geographers to be open to 'the study of geographical knowledge from any or all points of view ... [to] geographical ideas, both true and false, of all manner of people'. There are, I think, two important ways in which this claim to 'geosophy' played out historically. Earlier I focused on the commodification of culture and the ways in which such notions of 'local knowledge' merely extended the economy of display. But, in transformative mappings such as those by Bunge and Cravey, we see more clearly the progressive moment in Wright's claims for geosophy. Not only has a geosophic sensibility opened mapping to specific and different positionalities, but in so decentring the cartographic imagination, mapping practices have begun to pay more attention to the spaces of the everyday.

Marc Treib's 1980 monograph *'Mapping Experience'* reflects this concern with the many ways in which we do map everyday life. Focusing on the diversity and variety of mapped spaces, Treib sought to refocus attention on the ways in which cartographers were experimenting with new mapping forms to articulate experiences of space through new metrics and design features. *'Mapping Experience'* is a largely descriptive and evaluative document, short in length, and perhaps too assertive in tone. But it illustrates well the multiplicity and complexity of mapping forms that have emerged to chart the social cartography of spatial life. Treib's collection illustrates how these multiple spaces and forms have informed cartographic practice, and particularly how the city is *always already* being mapped in diverse ways, using a wide range of cartographic forms. In many ways, Treib's '*Mapping Experience*' symbolizes for me the diversity of cartographic experiments that followed 1968. The reworkings (and subsequent recommodification) of notions of subjectivity, experience and social life that so typified the revolutions of 1968, took root in the myriad cartographies of experience that were produced in its wake. If we look at mapping in this way (as already multiple, experimental, and open to flows, relations of difference, and change), we can, I think, begin to speak of cartographies as already and always involving imaginative open, contested and contradictory mappings.

SOCIAL CARTOGRAPHIES OF EXPERIENCE: GO ON, GO ON!

Thus, I have ended this book not with prescriptions for new techniques or practices of cartography, but with a question: what would cartography look like when we have overcome the modern settlement? Or, as Gibson-Graham might have asked, when we overcome the representational logics that bind us to a specific notion of cartography, what would cartography after Cartography look like? Or, again, when we abandon all forms of reduction and allow for the real possibility of logics of and ... and ... and ... and ... what kinds of cartography would be possible? How can we usefully and interestingly map 'lines of flight'? How are we *actually already* doing this even as we imagine and defend a rationalist and centred cartography?

Slavoj Zizek begins his book *The Ticklish Subject* (1999) with the mischievous question: what if, after all, Descartes was correct? What if, after all, we were to think of maps much as we have always thought about them? What if, after all, we were to continue to produce maps in much the same ways? In my writings on the political economy of post-communist transformation, I have taken great pains to stress the importance of focusing attention not on the categories that circulate so freely in communist and post-communist studies, but on actually occurring communisms and actually occurring transitions/transformations to capitalisms and to other forms of economic life. And, as here, I have expressed both as plurals: multiple communisms and transitions each at work across space and in places, each secreting their own spatialities and natures. In the final part of this book, then, I want to suggest that the axiomatization of modern thought, the abstraction of scientific–technological thinking, has developed an account of mapping, maps and cartography that belies the pragmatics of actual map-making and map use. It literally performs the God-trick on cartography's own lines of flight. That is, as Deleuze and Guattari have indicated in their analysis of the Oedipal fixation in psychoanalysis, and as Gibson-Graham (1996) have shown in *The End of Capitalism (as we know it)*, I am suggesting an 'end of cartography as we knew it' or that 'cartography is not what you think'. It is and perhaps has always been a multitude of practices ... lines of flight ... coded and recoded by forms of institutionalized power, but always with leakage. This decentring of the hegemonic formalization of techno-scientific capitalism opens mapping to its own plurality of socio-spatial practices, to its own geographies, to its own conflicted and highly contested nature, and to its many roles in inscribing lines and delimiting identities in the modern mind. Wittgenstein asked what would happen if, far off in the distance, the images began to oscillate? As Gunnar Olsson, Franco Farinelli and Marcus Doel have each recognized so well, our images and maps are already oscillating and shimmering. What has to begin to oscillate and shimmer more freely is our thinking about these actual practices.

None of this amounts to a call to re-historicize social life. I began this section with a discussion of the need to deepen the analysis of the taken-for-granted world and, in this context, I begin with Husserl, Heidegger and Wittgenstein in opposition to the historicizing traditions of Dilthey and the neo-Kantian historians. But there is another reason for avoiding the historicizing trap and it is stated strongly by Zizek (2001: 2):

> today's (late capitalist global market) social reality itself is dominated by what Marx referred to as the power of 'real abstraction': the circulation of Capital is the force of radical 'deterritorialization' (to use Deleuze's term) which, in its very functioning, actively ignores specific conditions and cannot be 'rooted' in them. It is no longer, as in the standard ideology, the universality that occludes the twist of its partiality, of its privileging a particular content; rather, it is the very attempt to locate particular roots that ideologically occludes the social reality of the reign of 'real abstraction'.

Since what Henri Lefebvre (1991) called the 1968 global–local crisis in social modernity, 'the production of space' has occurred in ways that have bound global and local, city and country, centre and periphery together in new and unfamiliar ways (Wilson and Dissanayake 1996: 3). Together these have fundamentally restructured the conceptual and institutional practices of mapping disciplines, and they are changing the ways in which we experience and understand earth, space and globality at this end/ beginning of century. New geographies have proliferated and these in turn have necessitated new categories and pedagogies.

This was, I think, what Fredric Jameson (1984: 89) was suggesting when he defined a new provisional aesthetic of cognitive mapping as one that places 'the analysis of representation on a higher and much more complex level'. Jameson (1984: 90) saw in the idea of cognitive map a parallel with the Althusserian and Lacanian redefinition of ideology as 'the representation of the subject's *Imaginary* relationship to his or her *Real* conditions of existence'. Jameson calls upon the cognitive map (and the social cartography it could produce) to 'enable a situational representation on the part of the individual subject to that vaster and properly unrepresentable totality which is the ensemble of the city's structure as a whole' (Jameson 1984: 90):

> An aesthetic of cognitive mapping – a pedagogical political culture which seeks to endow the individual subject with some new heightened sense of its place in the global system – will necessarily have to respect this now enormously complex representational dialectic and to invent radically new forms in order to do it justice. This is not, then, clearly a call for a return to some older kind of machinery, some older and more transparent national space, or some more traditional and reassuring perspectival or mimetic enclave: the new political art – if it is indeed possible at all – will have to hold to the truth of postmodernism, that is, to say, to its fundamental object – the world space of multinational capital – at the same time at which it achieves a breakthrough to some as yet unimaginable new mode of representing this last, in which we may again begin to grasp our positioning as individual and collective subjects and regain a capacity to act and struggle which is at present neutralized by our spatial as well as our social confusion. The political form of postmodernism, if there is any, will have as its vocation the invention and projection of a global cognitive mapping, on a social as well as a spatial scale.

In this sense, we can see the digital transition as a part of a broader post-Fordist development project; a global restructuring that is reconfiguring the geopolitics of the planet. The national and international imaginaries that emerged in an era of nation-state geopolitics are being reworked and new geo-political and geo-economic forms are emerging. Wilson and Dissanayake (1996: 2) have called this the 'process of translating the transnational structurations of nation, self, and community into 'translational', in-between spaces of negotiated language, borderland being, and bicultural ambivalence.' As a result 'The geopolitics of global cultural formations and local sites are shifting under the pressures of this new 'spatial dialectic' obtaining between mobile processes of transnationalization and strategies of localization or regional coalition.'

Beyond a political economy and geopolitics of technical change, a political technology of the social body and a corresponding regime of morality is also emerging, in which our understanding of the 'subject' itself is being reconfigured (see Hillis 1999a, 1999b; Uebel 1999). Felix Guattari (1991: 18) has called this 'the fabrication of new *assemblages* of enunciation, individual and collective' – in which actors and scales of action are no longer only governments and nation-states, but complex assemblages that go well beyond the military industrial complex of the 1950s and 1960s and multi-national corporations of the 1980s and 1990s. One way in which this is happening has to do with the very possibilities of the new technologies.

There is – as Michael Watts (n.d.) has written –

> a compelling paradox at the heart of globalization which turns on the differing ways in which material exchanges, forms of governance and authority, and symbolic interchange stand in relationship to place, territoriality or region. Globalization cannot simply be grasped as a solvent, or as an unalloyed force of cultural homogenization or geographical deterritorialization. For every instance of footloose financial services as a global space of flow and movement, there are other productive sectors characterized by economic rigidity and localization. For every case of the 'retreat of the state' there are equally compelling cases of enhanced state capacity. For every instance of global civil society or multilateral governance there are new configurations of national, local or regional politics. For every global technological or cultural diffusion, there is an equal and opposite intermixing and locally inventive appropriation. For every case of global cosmopolitanism and flexible citizenship there is a resurgence of local identity and 'militant particularism.' For every integrated global network there is, as Manuel Castells (1996) says, a black hole of displacement, exclusion and marginalization. Globalization seems to necessarily contain its opposite: the power of place and local identity, the ever-present local disjuncture and irruption, the multiplication of new forms of difference and heterogeneity.

For Watts (n.d., 1997) globalization is not displacing or undermining the importance of place or locale, but highlights the fact that much life is being conducted in 'globalized sites'. As Doreen Massey has so clearly shown, the flows, networks and movements that seem to be the hallmarks of globalization have not erased place or locality or region. First, globalization with its emphasis on the interactive world is not antithetical to the area – the region, the locality, the place, the nation – but reaffirms it in new and different ways. Second, globalization is an uneven, contradictory and complex set of processes perhaps best understood in quite specific 'globalized research sites'. Third, globalization challenges the classic notions of how we study and map the world at any scale, and calls for rethinking theory and method in 'globalized sites'. And fourth, globalization challenges the historic privileging of western cartographic logics and calls for rethinking and reconstructing mapping theory from more balanced comparative perspectives and materials.

What cartographies will be attentive to these rich respacings of social and political life? Overcoming the God-trick means paying much more attention to the multiplicity and diversity in what previously passed for unity. It means deconstructing and disseminating both traditional Cartesian anxieties and the anxieties that see in maps only instruments of power. But it must also see in this analytics of complexity something other than a merely liberal reading of benign technologies and instruments put to good or bad uses (see Monmonier 2002). The openness to difference is a much more radical epistemological opening of the sutured politics of contemporary cartographics. Such new cartographies might deploy every technical tool to produce mappings that speak their situated and selective interestedness, that record their metadata and political commitments, and that recognize the pragmatic nature of their own practice. But it is also a cartography that needs a new openness to producing dialectical, dynamic and metaphorical images; one that resists collapsing striated to smooth space, the local to the global, or the concrete and particular to the abstract and universal. It is, above all, a cartography that would be attentive to the serious consequences of the lines we draw and the boundaries we inscribe in the very broadest of terms (Deleuze 1988).

In a series of reflections on the cartographies of borders, the changing nature of citizenship, the shifting relationship between *ethnos* and *demos* in the twinned 'nation-state', and the post-national order of Europe, Etienne Balibar (2002a, 2002b) has recently focused on precisely such dialectical cartographies of geographical transformation and on what he calls the 'vacillating' nature of contemporary borders (2002a: 91). For Balibar (2002b: 71) the borders of new politico-economic entities are no longer localizable in an unequivocal fashion, nor are they situated only (or at all) at the outer limit of territories. They are not disappearing under the pressure of globalization. Instead, they are being multiplied, thinned out and doubled: they are 'dispersed a little everywhere': to the outer limits of the European Union, to the Schengen signatory states, to the inner limits of cosmopolitan cities. The reordering of citizenship and civic rights in the globalized modern state redraws the border, and its mark is carried with the immigrant daily. In this sense 'border areas are not marginal to the constitution of a public sphere but rather are at the centre' (Balibar 2002b: 72).

Contemporary globalization brings with it what Balibar (2002a: 93) calls an 'under-determination of the border' and a dispersal and proliferation of their roles in defining citizenship, forms of inclusion and exclusion, policing, and identification. In this sense, every map is always a 'world' map and in this changing world we need new cartographies that evoke the vacillating, dispersed and disseminated nature of borders. The cartographies that emerged to 'service' the territorialized nation-state of an earlier round of globalization – Europe as the universal centre of politics, thought and economy – defined the global by universalizing largely European values. As Chakrabarty (2000: 4) suggests:

> The phenomenon of 'political modernity' – namely, the rule by modern institutions of the state, bureaucracy, and capitalist enterprise – is impossible to *think* of anywhere in the world without invoking certain categories and concepts, the genealogies of which go deep into the intellectual and even the theological traditions of Europe. Concepts such as citizenship, the state, civil society, public sphere, human rights, equality before the law, the individual, distinctions between public and private, the idea of the subject, democracy, popular sovereignty, social justice, scientific rationality, and so on all bear the burden of European thought and history ... These concepts entail an unavoidable – and in a sense indispensable – universal and secular vision of the human ... [which] has been powerful in its effects.

Balibar reminds us of the need to see always in our inscriptions forms of boundary-making that have effects. In the contemporary world of globalizing transnationalisms the boundaries and borders that shape and structure the geographies of inclusion and exclusion, property and citizenship, *ethnos* and *demos* require new cartographies of geographies unhinged, plastic space and sliding signs (Doel 1999). We need new diagrams, abstract machines and maps that are attentive to these highly differentiated reconfigurations of time and space, and to the new notions of nationhood, citizenship, state and territory they entail.

It is here that we again encounter Gunnar Olsson's continued explorations with the cartography of power: 'No rest, no escape. GO ON, GO ON! The explorations into the taken-for-granted must continue' (Olsson 1994: 115). How are we to 'draw the invisible lines of the taken-for-granted?' How are we to speak so that we are understood, to say that something is something else and still be believed? This, indeed, is the trick of the magician, the poet and the scientist. It is the goal of cartographic imagination.

Drawing on Olsson's arguments, the Italian geographer Franco Farinelli (1999) has called for a geography that recognizes the 'Witz' or joke or witticism of 'bat-words' (mouse/bird) like landscape, space, world, earth; words that contain what Olsson calls an ambiguous duplicity of meanings ... at once material, artistic, ideation, and lived. If the epistemology of modernity fixes meaning, the emergence of epistemologies and mappings of transparency open up the possibility of thinking about the world-not-as-picture and the world-not-as-exhibition, but in terms of new dialectical images that render movement as movement, rather than frozen images, dead, inert, fixed.

The challenge ahead then is precisely the challenge with which we began. How can we open our everyday and professional practices to new cartographies and new geographies? I end with three answers.

First, as Bill Bunge has demonstrated so well, our existing cartographies and categories are far less fettered than we have perhaps acknowledged. This is not to say that traditional and contemporary cartographies have always been, or are currently open to these new cartographies. It is to say that it may be possible to develop new cartographies and geographies only by changing the way we think about the cartographies we have. The end of cartography as we know it is, as Gibson-Graham, Deleuze and Guattari, and Negri and Hardt have variously shown us, the possibility of opening the contradictory moments within existing practices to new opportunities and alternative projects.

Second, experiments with shattered logics, flowing art forms, and situational performance are highly productive and suggestive. They expand in important ways both our practices of mapping and our imaginations about the 'Real', and they do so in ways that destabilize all forms of the God-trick, universalism and the march of progress. They force us to understand the pragmatics of map use and the social embeddedness of map-making. In such perspectives, the mapping sciences can usefully be reconnected to the actual practices of what has always been a fractal cartography of complexity. No longer a cartography of statecraft, of the centred and nominally universal polity, but a cartography of ongoing space–time reconfigurations; new boundary-making always with potentially serious consequences.

Finally, if the new cartographies are already with us, we must also recognize that they do not have a unitary and fixed identity. The abstractive mappings of von Humboldt's planetary consciousness, the progressive struggles of spatial analysis, the conceptual flexibilities and political possibilities of the Digital Earth Initiative and the *Atlas of Cyberspaces* have already de-ontologized whatever we ever meant by modern cartography in ways that we are perhaps only beginning to recognize. In this sense, Foucault will always be correct when he claimed that a whole history of spaces remains to be written.

Notes

1 Maps and worlds

1 'On persuasion and power', presentation to the Committee on Social Theory, University of Kentucky, 29 March 1991.
2 In this new world of images, commodity fetishes and dream fetishes become indistinguishable. Food and other commodities drop magically onto the shelves of stores, and advertising and commerce come to be seen as the means of social progress. The democratization of culture is now seen to derive from the mass media, and they too become fetishes (Buck-Morss 1989: 120).
3 The intimacy of this perceived relationship is all too clearly illustrated in Hartshorne's (1939: 248) quotation from P.E. James: 'The most important contributions of geography to the world's knowledge have come from an application of the technique of mapping distributions and of comparing and generalizing the patterns of distributions'.
4 Gregory (1994) used the term 'Cartographic Anxiety' to refer to the foundational and objectivist epistemologies of modern cartography that assume the separation of subject and object, knower and world. This 'observer epistemology' leads to deep anxiety about how we know and represent the world, how we know it to be true, and how we decide what to do in the face of such 'objective' knowledge. The term is adapted from Richard Bernstein's use of 'Cartesian Anxiety' in *Beyond Objectivism and Relativism: Science, Hermeneutics and Praxis*. This anxiety refers to what Bernstein (1983: 18) calls: 'The specter that hovers in the background ... not just radical epistemological skepticism but the dread madness and chaos where nothing is fixed, where we can neither touch bottom nor support ourselves on the surface. With a chilling clarity Descartes leads us with an apparent and ineluctable necessity to a grand and seductive Either/Or. *Either* there is some support for our being, a fixed foundation for our knowledge, *or* we cannot escape the forces of darkness that envelop us with madness, with intellectual and moral chaos'.
5 In a similar way, Cosgrove's (1999: 1) recent collection of essays *Mappings* focuses on: 'the long evolution of western spatiality in order to explore some of the contexts and contingencies which have helped shape acts of visualizing, conceptualizing, recording, representing and creating spaces graphically – in short, acts of *mapping*'.
6 I use the term 'dissemination' in a Derridean sense to refer to all the ways in which we can see at work in mapping practices, multiple epistemological and geographical visual regimes.
7 Instead of using the published form of this table (see Woodward and Lewis 1998 Table 1.1), I have retained its pre-publication form kindly supplied to me by David Woodward. The published table reworks the categories of process

(thought and performance) and product (record/material cartography) into internal/inner experience (cognitive cartography) and external (performance and material cartography). This inscription of a Cartesian inner–outer, cognitive–material distinction sits uncomfortably with the richer, more nuanced text and its careful treatment of traditional mapping forms, and runs counter to the argument I make here.

8 Berry's *Essays on Commodity Flows and the Spatial Structure of the Indian Economy* contains 101 maps of the Indian subcontinent, including an atlas of Indian commodity flows. The maps are sequenced one after another building a composite image that both illustrates and models (produces/defines/delimits) the abstract spaces of interaction and structure.

9 Reading contrapuntally involves 'a simultaneous awareness both of the metropolitan history that is narrated and of those other histories against which (and together with which) the dominating discourse acts' (Said 1999: 51). The musical metaphor has been adapted by Sparke (1998: 467) to read against singularized and unidirectional accounts of imperialist cartographies.

10 It is, of course, from Foucault that I have taken both the title and the inspiration for this book. In this sense *A History of Spaces* represents the third part of a threefold engagement with the philosophy and geo-history of spatiality. The first part dealt with the hermeneutic ontology of spatiality and was laid out in *Phenomenology, Science and Geography* (1985). The second part focused on the geopolitics of socio-spatial life in societies undergoing rapid transformation, and was published variously in *Theorizing Transition* (1988), *Bulgaria in Transition* (1998), and *Environmental Transitions*. This third part marks the fuller consideration of the space–power and power–knowledge of spatial practices signalled in the final chapter of *Phenomenology, Science and Geography*.

2 What do maps represent? The crisis of representation and the critique of cartographic reason

1 At least four streams of this work need to be mentioned: (a) the construction of modern categories of socio-spatial identity, e.g., Martin Bulmer, Kevin Bales and Kathryn Kish Sklar (1991); (b) the growing body of work on the role of mapping in the construction of national identity, e.g., Thongchai (1994), Anderson (1991) and Krishna (1996); (c) the related work on mapping and the imperial project, e.g., Pratt (1992) and Godlewska, (1995); and (d) the metaphorical deployment of mapping in the context of critical social theory and geopolitics, e.g., Pile and Thrift (1995) and Shapiro (1997).

2 *Social Cartography* is a product of the Social Cartography Project run out of the University of Pittsburgh's School of Education, Department of Administrative and Policy Studies Department during 1993–6, and (the editor claims) represents the spectrum of international work dealing with ideational mapping. The collection is organized in four sections, each with its own brief introduction, and each containing four to six papers (I 'Mapping imagination', II 'Mapping perspectives', III 'Mapping pragmatics' and IV 'Mapping debates'). The range of papers is wide and they are correspondingly diverse in content and approach: from modern to postmodern ways of seeing social and educational change, 'knowledge spaces' and sites of resistance, the origins of social cartography, spatial analysis, mythopoeic images, strategic thought, spatial metaphors, mapping gendered spaces, utopias, rural development, intercultural communication in educational consultancies, environmental education discourses, perception, 'subalternity', post-colonial feminism, the spaces of capital, Jameson's cognitive mapping and critical modernism.

3 Here I am thinking of Denis Wood's (1992) *The Power of Maps*, but also the concept of power at work in Brian Harley's (1988, 1989a, 1989b) conception of maps. In Harley's work, 'deconstruction' also seems to function as a form of ideology critique in which particular interests and effects of power are to be unmasked.

4 I italicize *transparency* here to indicate a difference/distinction that will emerge in later chapters where I turn to new mapping ontologies of transparency.

5 Krishna (1996: 194) captures it this way: 'By cartography I mean more than the technical and scientific mapping of the country. I use the term to refer to representational practices that in various ways have attempted to inscribe something called "India" and endow that entity with a content, a history, a meaning, and a trajectory. Under such a definition, cartography becomes nothing less than the social and political production of nationality itself.'

6 Derek Gregory (1994) has referred to the goals of this modern mapping impulse as systematicity, boundedness and totalization.

7 The interested reader might wish to turn to Chapters 1–3 of *Phenomenology, Science and Geography*, where I addressed this same crisis through a wider-ranging critique of objectivism and subjectivism in geography.

8 For a broader discussion of the contested nature of maps and the appropriate approaches to conceptualizing them, see Downs and Stea (1973, 1977).

9 Perhaps no clearer illustration of this abiding commitment to Cartesian and Kantian epistemologies can be found than that in Pequet's recent (2002: 33) *Representations of Space and Time*. Here representation and reality are again 'set up' as oppositional categories and questions are posed about how to resolve the antinomies in ways very much like those J.K. Wright grappled with fifty years ago: 'What is the nature of the process involved in gaining knowledge of the external world and construction of a "world-view"? What are the commonalities in internal world-views, and how do these relate to commonalities in representations of geographic space? How do these internal representations relate to external representations used to communicate that knowledge.'

10 I am grateful to an anonymous reviewer for emphasizing that, 'For much of his career, Harley, like most cartographers, was a firm believer in the map as mirror, objectivity, and the aspiration toward error-free maps. Even as late as 1980 when he published his "Concepts in the History of Cartography: A Review and Perspective", co-authored with Mike Blakemore (*Cartographica* monograph 26 v 17(4)), Brian was trying to elaborate a typology of error . . . By the late 1980s he had dropped these ideas.'

3 Situated pragmatics: maps and mapping as social practice

1 Nomadism seems to capture well the apparent ease with which Wood's cartographies of experience travel from one domain of everyday life to another. See Wood (1977a, 1977b, 1978a, 1978b, 1980).

2 This paragraph is a paraphrase and elaboration of Bijker and Law's (1992: 1).

3 I am grateful to Francis Harvey for bringing this ecological metaphor to my attention.

4 For parallel discussions dealing explicitly with geographical information systems see the following: GIS–social theory debates (Mark *et al.* n.d.; Pickles 2000a), the social implications of GIS (Pickles 1991, 1995), GIS and the politics of knowledge (Pickles 1993, 1997), the political economy of GIS adoption in Eastern Europe (Pickles and Mikhova 1988).

4 The cartographic gaze, global visions and modalities of visual culture

1 The current role of mapping as a metaphor also signals the strength of this kind of cartographic reason in social life and thought. See, for example, the *Mapping* series (e.g., Zizek 1995; Balakrishnan 1996).
2 Rose evokes Lucy Irigaray's work on phallocentrism, the mirror image and the space of self-knowledge, through which she discusses the ways in which men dominate the representational economy and in which difference – far from being a source of alternative, contingent knowledge claims and understandings of the world – becomes instead a negative of itself, the inverted other of the masculine subject.
3 Lestringant even suggests that 'In their very excess, Thevet's incessant marginal grafts and captions define what could be called a rhetoric of plagiarism' (Lestringant 1994: 128).

5 Cadastres and capitalisms: the emergence of a new map consciousness

1 I am, of course, aware that some readers may see me conflating the ideas of authors as diverse as Derrida, Borges, Baudrillard and Lefebvre. But in this instance I am interested in the ways in which each denaturalizes social space, territory and identity. More generally, as I indicated in the Preface, I seek a more open hermeneutic reading of critical social theory, one that strives to sharpen analytical distinctions but is also open to commonalities in difference. I am, that is, interested in the productive potential of border crossings more than I am in border policing.
2 The *Oxford English Dictionary* defines a 'cadastre' as: '**a.** The register of *capita* ... or units of territorial taxation into which the Roman provinces were divided for the purposes of a ... land tax ... **b.** A register of property to serve as a basis of proportional taxation ... **c.** (in modern French use) A public register of the quantity, value, and ownership of the real property of a country' and a 'cadastral survey' as '**b.** a survey on a scale sufficiently large to show accurately the extent and measurement of every field and other plot of land'.
3 Jowett *et al.* (1992: 205). 'Among the questions Turgot [Louis XVI's controller general of finances] had to address in the latter part of the eighteenth century was whether the monarchy could ever again achieve financial solvency and whether government could deploy the power to "make the public interest prevail over the constitutional prerogatives of the privileged orders and corporate bodies, ending or abating their exemption from direct taxation"' from Charles Coulston Gillispie (1980: 4); and 'It was for an instrument to effect this latter that the French state turned to cadastral mapping' Marc Bloch (1929) 'Les Plans parcellaires' (p. 392).

6 Mapping the geo-body: state, territory and nation

1 This was, of course, never an argument that GIS and mapping were merely war technologies. Practitioners in geography were certainly involved directly and indirectly in the development of smart-bombing technologies and many of the technologies and practices we use in GIS and computerized cartography have been underwritten by military and covert intelligence research within military research establishments (for example, human-interface and visualization technologies) or through public and secret funding of university research (see, for example, the case of TIN research in GIS (Mark 1997)).

7 Commodity and control: technologies of the social body

1 I am grateful to Jim Hevia for bringing to my attention this essay and Holmes's interest in stereoscopes.

2 It is clear in many of these 'confessionals' that the authors had a strong sense that a new field of thought was being opened up by their encounter with a cartographic imagination. Many of them go on to recount how their later experiments in spatial abstraction, which changed the categories and analytics through which the world was structured, were influenced by this cartographic experience.

3 For example, Cesare de Seta (1994b: 126) locates the 'new passion of the time – sea bathing' from the early to mid-nineteenth century.

4 In the concluding chapter I revisit the notion of 'counter-images' as a way of rereading the significance of Bill Bunge's cartographic project precisely as 'counter-mappings' in a Benjaminian sense.

8 Cyber-empires and the new cultural politics of digital spaces

1 These technological embodiments and penetrations of everyday life are never more apparent and sharp than at times of modern war.

2 For a discussion of the political economy and cultural politics of the notion of a 'digital transition', see Goodchild (2000), Pickles (2002), and Rhind (2000). See also Evans and Leder (1999) where a standard evolutionary history of socio-technical change is described: 'The first industrial revolution lasted from 1760 to 1850 and was responsible for widespread innovations ranging from steam engines to iron production. Between 1890 and 1930, the second industrial revolution brought us electricity, telephones and the internal combustion engine. Today's industrial revolution stems from two innovations from the second half of the 20th century. First, a rudimentary international computer network (later to be called the Internet) and second, the development of the world's first microcomputers (better known today as PCs)'. http://www.mmc.com/views/99sum.evans.shtml

3 Throughout this discussion I am attempting to tease out some of the cultural imaginaries and unstated assumptions and commitments within which contemporary cartographic practices are located. I do not attempt here an imminent analysis of the cartographic imagination in use in geographical information systems.

4 This view is foreshadowed in negative critical form as the 'totally administered society' in Horkheimer and Adorno's (1944) *The Dialectic of Enlightenment*. See also Kellner (1989: 98–9).

5 Part of this wider tradition includes multimedia and hypertext. Mark Poster (1990) has perhaps provided the most thorough theorization of the new revolution in visualization brought about by new electronic information systems, but it is in the work of Landow (1992) that poststructuralist ideas are brought directly to bear on an interpretation of multimedia and hypertext. For Landow (1992: 2) critical social theory promises a way of theorizing hypertext, and hypertext embodies and tests theories of textuality, narrative, margins, inter-textuality and the roles and functions of readers and writers. In Roland Barthes's term, hypertext produces writerly texts that do not dominate the reader and insist on particular readings, but engage the reader as an 'author' and insist upon the openness and intertextuality of the text – that is, its open-ness to other texts and readings. 'When designers of computer software examine the pages of *Glas* or *Of Grammatology*, they encounter a digitalized, hypertextual Derrida; and when literary theorists examine *Literary Machines*,

they encounter a deconstructionist or poststructuralist Nelson. These shocks of recognition can occur because over the past several decades literary theory and computer hypertext, apparently unconnected areas of inquiry, have increasingly converged.'

6 Of course, all these matters are contingent on the types of regulatory framework that emerge to govern development, property rights, access, etc. For further discussion of this point, see Pickles and Mikhova (1998) and Pavlinek and Pickles (2000).

7 The primary information sector refers to computer manufacturing, telecommunications, mass media, advertising, publishing, accounting, education, research and development, and risk management in finance, banking, and insurance. The secondary information sector refers to work performed by information workers in government and goods-producing and service-producing firms for internal consumption (Luke and White 1985: 33).

8 I am not suggesting that this situation is unique to mapping sciences or that any links are simple or straightforward. In some ways, the issue is generic to all those techno-sciences that have developed historically in conjunction with the practices of statecraft. In these cases, too, the institutional connections are complex, at times conflicted, and often ambiguous.

9 Doug Kellner (1989: 178) uses 'technocapitalism' as the term to describe 'a configuration of capitalist society in which technical and scientific knowledge, automation, computers and advanced technology play a role in the process of production parallel to the role of human labor power, mechanization and machines in earlier eras of capitalism, while producing as well new modes of societal organization and forms of culture and everyday life'.

10 Benjamin (1968) suggested that the new often enters onto the stage of history presented in the guise of its predecessor. Thus, the train emerges in the guise of a roaring bull, and the computer emerges as, and in the form of, a typewriter or a counting machine.

11 I use Stenger's claims with caution. The underlying conception of social action in Stenger is highly problematic. Indeed it is, I think, antithetical to that being developed here and that suggested by her own definition of guerrilla epistemology: 'But the guerrilla has to imagine himself [sic] as belonging to a disciplined army, and relate the sense and possibility of his local initiatives to the commands of staff headquarters.'

12 http://www.cybergeography.org/atlas/atlas.html

13 See the US Institute of Health's Visible Human Project: http://www.nlm. nih.gov/research/visible/visible_human.html; http://www.nlm.nih.gov/research/ visible/visible_gallery.html; http://www.madsci.org/~lynn/VH/planes.html

14 The first visible man used to produce the 'Visible Human' in what Catherine Waldby (2000) calls post-human medicine was, of course, neither chimera nor ghost but a very real person. He was condemned murderer, Joseph Paul Jernigan, executed in Texas. He was joined two years later by a 'Maryland housewife' who had died of a heart attack, the first 'Visible Woman' (Waldby 2000: 1). While the 'Visible Human Project' brings the structure of the human body so clearly into view it does so by fetishizing the visible and erasing the material bodies of the two invisible subjects whose body was sliced and diced into the thousands of thin sections in order to produce the 'Visible Human'. I am grateful to an anonymous reviewer for this particularly graphic 'rendering'.

15 Sources for Digital Earth include: http://digitalearth.gsfc.nasa.gov/; http:// www.digitalearth.gov/; http://www.digitalearth.gov/vision.html; http://www. icase.edu/~tom/DigitalEarth/DEResources.html.

16 See also the Global mapping project at http://www1.gsi-mc.go.jp/iscgm-sec/

17 Lefebvre (1991: 31–3) argued that every society secretes its own spaces. It is this notion of societies secreting their own spaces that is intended here.
18 For a parallel argument, see Ould-Mey (1994: 319–36).
19 These few paragraphs on digitality and intertextuality rely on the themes developed in more detail in George Landow's (1992) *Hypertext: The Convergence of Contemporary Critical Theory and Technology*. For a discussion of the notion of text implied here, see Derrida (1986: 366): '*text*, as I use the word, is not the book. No more than writing or trace, it is not limited to the *paper* which you cover with your graphism. It is precisely for strategic reasons … that I found it necessary to recast the concept of text by generalizing it almost without limit, in any case without present or perceptible limit, without any limit that *is*. That's why there is nothing "*beyond* the text".'

9 Counter-mapping: cartographic reason in the age of intelligent machines and smart bombs

1 In *Phenomenology, Science and Geography: Space and the Human Sciences*, I argued that Heidegger's analytic of finitude in *Being and Time* and subsequent works was explicitly responding to this need to incorporate an analytic of spatiality in contemporary philosophy.
2 In considering Bunge, one might see a nostalgia for 1960s and early 1970s politics of engagement or, as one reviewer of this manuscript suggested: 'I'm amazed that there the author is engaging in an unabashed "return to the 1960s"'. This may well be true. But I also want to suggest that we can read Bunge's 'counter-mappings' in ways that bring them closer to Benjamin's 'counter-images'; dialectical images that both reflect different interests and perspectives, but also that thoroughly deconstruct the metaphysics of presence, epistemology of representation and politics of progress on which the universality projects of cartography and GIS rest so heavily and so comfortably.

References

Aberley, D. (ed.) 1993. *Boundaries of Home: Mapping for Local Empowerment*, Philadelphia, PA: New Society Publishers.

Abler, R.A. 1993. 'Everything in its place: GPS, GIS and geography in the 1990s', *Professional Geographer*, 45(2): 131–9.

Abler, R.F., Adams, J.S. and Gould, P.R. 1971. *Spatial Organization: The Geographer's View of the World*, Englewood Cliffs, NJ: Prentice Hall.

Adorno, T.W. 1973. *Negative Dialectics*, New York, Continuum: 139–40.

Adorno, T.W. 1991. 'Transparencies on film', in J.M. Bernstein (ed.), *The Culture Industry: Selected Essays on Mass Culture*, London: Routledge.

Ager, J. 1977. 'Maps and propaganda', *Bulletin of the Society of University Cartographers*, 11: 1–15.

Agger, B. 1989. *Fast Capitalism: A Critical Theory of Significance*, Urbana and Chicago: University of Illinois Press.

Agnew, J. 1998. *Geopolitics: Re-visioning World Politics*, London and New York: Routledge.

Alpers, S. 1983. *The Art of Describing: Dutch Art in the Seventeenth Century*, Chicago: Chicago University Press.

Amin, S. 1994. *Re-Reading the Postwar Period: An Intellectual Itinerary*, New York: Monthly Review Press.

Anderson, B. 1991. *Imagined Communities: Reflections on the Origins and Spread of Nationalism*, London: Verso.

Appadurai, A. 1996. *Modernity at Large*, Minnesota: University of Minnesota Press.

Appadurai, A. 1997. 'The research ethic and the spirit of internationalism', *Items (SSRC)*, 51(4:1), December 1991: 55–60.

Aster, E.D. 1996. Publisher's letter, *Mercator's World*, 1(1): 7.

Aziz, B.N. 1978. 'Maps and the mind', *Human Nature*, 1(8), August: 50–9.

Bachelard, G. 1984. *The New Scientific Spirit*, New York: Beacon.

Bacon, F. 1605. *The Advancement of Learning*. www1.uni_bremen.de/~kr538/baconadv.html.

Balchin, W.G.V. and Coleman, A.M. 1965/1966. 'Geographicacy should be the fourth ace in the pack', *Cartographer* 3(1), June: 23–8.

Balakrishnan, G. (ed.) 1996. *Mapping the Nation*, London and New York: Verso.

Balibar, E. 2002a. *Politics and the Other Scene*, London: Verso.

Balibar, E. 2002b. 'World borders, political borders', *PMLA*, 117/1, January, 71–8.

Barnes, T. 1981. 'On geography and cartography', Discussion Paper 5, Department of Geography, University of Minnesota. Mimeo 4 pp.

Barry, D. 1991. 'Fleet management makes advances with digital mapping technology', *GIS World*, October: 74–7.

Barry, J. 1992. 'Mappings – a chronology of remote sensing', in J. Crary and S. Kwinter (eds), *Incorporations* Zone 6, Cambridge, MA: MIT Press: 570–1.

Barthes, R. 1978. *Image, Music, Text*, New York: Noonday.

Bassett, T.J. 1994. 'Cartography and empire building in nineteenth-century West Africa', *Geographical Review*, 84(3): 316.

Baudrillard, J. 1981. *For a Critique of the Political Economy of the Sign*, St Louis, Missouri: Telos.

Baudrillard, J. 1983. *Simulations*, trans. Paul Foss, Paul Patton and Philip Beitchman, New York: Semiotext(e).

Baudrillard, J. 1988. 'The consumer society', in M. Poster (ed.), *Jean Baudrillard: Selected Writings*, Stanford: Stanford University Press.

Baudrillard, J. 1994. 'The system of collecting', in J. Elsner and R. Cardinal (eds), *The Cultures of Collecting*, Cambridge, MA: Harvard University Press: 7–24.

Bauman, Z. 1992. *Intimations of Postmodernity*, London and New York: Routledge.

Belyea, B. 1992. 'Images of power: Derrida/Foucault/Harley', *Cartographica*, 29(2), summer: 1–9.

Benedikt, M. 1991. *Cyberspace: First Steps*, Cambridge, MA: MIT Press.

Benjamin, W. 1968. 'The work of art in the age of mechanical reproduction', in H. Arendt (ed.), *Illuminations: Essays and Reflections*, New York: Schocken Books: 217–57.

Benjamin, W. 1999. *The Arcades Project*, Cambridge, MA: Belknap Press (an imprint of Harvard University Press).

Berger, J. 1965. *Ways of Seeing*, London and Harmondsworth: British Broadcasting Corporation and Penguin.

Bernal, J.D. 1971. *Science in History*, Cambridge, MA: MIT Press.

Bernstein, R.J. 1983. *Beyond Objectivism and Relativism: Sciences, Hermeneutics and Praxis*, Philadelphia: University of Pennsylvania Press.

Berry, B.J.L. 1966. *Essays on the Commodity Flows and the Spatial Structure of the Indian Economy*, Department of Geography Research Paper 111, Chicago: The University of Chicago.

Bijker, W.E. and Law, J. 1992. *Shaping Technology/Building Society. Studies in Sociotechnical Change*, Cambridge, MA: MIT Press.

Blachut, T.J. and Burkhardt, R. 1989. *Historical Development of Photogrammetric Methods and Instruments*, Falls Church, VA: American Society for Photogrammetry and Remote Sensing.

Black, J. 2000. *Maps and History: Constructing Images of the Past*, New Haven and London: Yale University Press.

Black, J. 2002. *Maps and Politics*, London: Reaktion Books.

Blakemore, M. 1981. 'From way-finding to map-making: the spatial information fields of aboriginal peoples', *Progress in Human Geography*, 5(1): 1–24.

Blakemore, M. 1990. 'Cartography', *Progress in Human Geography*, 14(1): 101–11.

Blakemore, M. 1992. 'Cartography', *Progress in Human Geography*, 16(1): 75–87.

Blaut, J. 1991. 'Natural mapping', *Transactions of the Institute of British Geographers*, NS, 16: 55–74.

Blaut, J. 1993. *The Colonizer's Model of the World*, New York: The Guilford Press.

Bloch, M. 1929. 'Les plans parcellaires', *Annales d'Histoire Economique et Sociale* 1: 60–70, 390–98, 392.

Board, C. 1967. 'Maps as models', in R.J. Chorley and P. Haggett (eds), *Models in Geography*, London: Methuen: 671–726.

Boggs, S.W. 1947. 'Cartohypnosis', *Scientific Monthly*: 469–76.

Bohlen, C.E. 1973. *Witness to History, 1929–1969*, New York: Norton.

Booth, C. 1902. *Life and Labour of the People in London, Volume 1*, Macmillan: London.

Borges, J.L. 1964. 'Museum: on rigor in science', in *Dreamtigers*, trans. Mildred Boyer and Harold Morland, Austin: University of Texas Press.

Boxer, C.R. 1969. *The Portuguese Seaborne Empire, 1415–1825*, London: Hutchinson.

Brealey, K.G. 'Networks of power: cartography as ideology', M.A. thesis, Department of Geography, Simon Fraser University, Burnaby, British Columbia.

Broek, J.O.M. 1965. *Geography: Its Scope and Spirit*, Columbus, OH: Charles E. Merrill.

Brown, L.A. 1969. *The Story of Maps*, New York: Bonanza Books.

Browning, C.E. 1982. *Conversations with Geographers: Career Pathways and Research Styles*, Studies in Geography 16, Department of Geography, University of North Carolina at Chapel Hill.

Buck-Morss, S. 1989. *The Dialectics of Seeing: Walter Benjamin and the Arcades Project*, Cambridge, MA: MIT Press.

Buisseret, D. 1988. *Rural Image: The Estate Plan in the Old and New Worlds*, Chicago: The Newberry Library.

Buisseret, D. (ed.) 1992. *Monarchs, Ministers and Maps*, Chicago: University of Chicago Press.

Buisseret, D. and Akerman, J.R. 1995. *Maps, Ministers and Monarchs: A Cartographic Exhibit at the Newberry Library on the Occasion of the Eighth Series of Kenneth Nebenzahl Jr. Lectures in the History of Cartography*, Chicago: The Newberry Library.

Bulmer, M., Bales, K. and Sklar, K.K. (eds) 1991. *The Social Survey in Historical Perspective 1880–1940*, Cambridge, MA: Cambridge University Press.

Bunge, W. n.d. *Geographical Prejudices*. Mimeo.

Bunge, W. 1962. *Theoretical Geography*, Lund Studies in Geography, Lund: Gleerup.

Bunge, W. 1969. 'The first years of the Detroit Geographical Expedition: a personal report', *Field Notes* no. 1: 1–9, reproduced in R. Peet, *Radical Geography*, Chicago: Maroufa Press: 31–9.

Bunge, W. 1971. *Fitzgerald: Geography of a Revolution*, Cambridge, MA: Schenkman.

Bunge, W. 1975. 'Detroit humanly viewed: the American urban present', in R.A. Abler, D. Janelle and A. Philbrick (eds), *Human Geography in a Shrinking World*, North Scituate, MA: Duxbury Press: 149–81.

Bunge, W. 1988. *Ban the Bomb: The Nuclear War Atlas*, Oxford: Blackwell.

Burnett, A. 1985. 'Propaganda cartography', in D. Pepper and A. Jenkins (eds), *The Geography of Peace and War*, New York and Oxford: Blackwell: 60–88.

Carroll, L. 1894. *Sylvie and Bruno Concluded*, New York and London: Macmillan.

Castells, M. 1996. *The Rise of the Network Society*, Oxford: Blackwell.

Chakrabarty, D. 2000. *Provincializing Europe: Postcolonial Thought and Historical Difference*, Princeton, NJ: Princeton University Press.

Chamberlain, P. 2001. 'The Shakespearean globe: geometry, optics, spectacle', *Environment and Planning D: Society and Space*, 19(3), June, 317–33.

Ciccone, D.S., Landee, B. and Weltman, G. 1978. *Use of Computer-generated Movie Maps to Improve Tactical Map Performance*, Technical Report PTR-1033-S-78-4, Perceptronics, Inc., Woodland Hills, CA.

Clark, G.L. 1992. 'Commentary: GIS – what crisis?', *Environment and Planning A.*, 321–2.

Clark, M. 1987. *The World is Found: Contemporary Panoramas*, New York: The Hudson River Museum.

Clark, R.E. 1991. 'GIS solves Denver car dealer's dilemma', *GIS World*, October: 67–73.

Clarke, K.C. 1986. 'Advances in geographic information systems', *Computers, Environment and Urban Systems*, 10: 175–84.

Clarke, K.C. 1992. 'Maps and mapping technologies of the Persian Gulf War', *Cartography and Geographic Information Systems*, 19(2): 80–7.

Clarke, S. 1988. 'Overaccumulation, class struggle and the regulation approach', *Capital and Class*, 36: 59–92.

Cohen, J.L. and Arato, A. 1992. *Civil Society and Political Theory*, Cambridge, MA: MIT Press.

Conley, T. 1996. *The Self-Made Map: Cartographic Writing in Early Modern France*, Minneapolis: University of Minnesota Press.

Cooke, D.F. and Plumer, C. 1991. *GIS World*, October: 82–5.

Coppock, J.T. and Rhind, D.W. 1991. 'The history of GIS', in D.J. Maguire, M.F. Goodchild and D.W. Rhind (eds), *Geographical Information Systems*, 21–43, London: Longman Scientific and Technical; New York: John Wiley & Sons, 2 vols.

Cosgrove, D. 1988. 'Prospect, perspective, and the evolution of the landscape idea', *Transactions of the Institute of British Geographers*, NS, 10: 45–62.

Cosgrove, D. 1989. 'Looking in on our world: images of global geography', in P. Wombell (ed.), *The Globe: Representing the World*, York: Impressions: 13–18.

Cosgrove, D. 1994. 'Contested global visions: *One-World, Whole-Earth*, and the Apollo Space photographs', *Annals of the Association of American Geographers*, 84(2): 270–94.

Cosgrove, D. 1998. *Social Formation and Symbolic Landscape*, Madison: University of Wisconsin Press, 2nd edn.

Cosgrove, D. (ed.) 1999. *Mappings*, London: Reaktion Books.

Cosgrove, D. 2000. 'Global illumination and enlightenment in the geographies of Vicenzo Coronelli and Athanasius Kircher', in C. Withers and D. Livingstone (eds), *Enlightenment Geographies*, Chicago: Chicago University Press: 33–66.

Cosgrove, D. 2001. *Apollo's Eye: A Cartographic Genealogy of the Earth in the Western Imagination*, Baltimore and London: The Johns Hopkins University Press.

Couclelis, H. 1988. 'The truth seekers: geographers in search of the human world', in R.G. Golledge, H. Couclelis and P. Gould (eds), *A Ground for Common Search*, Santa Barbara: Santa Barbara Geographical Press: 148–55.

Craik, K.H. 1977. 'Environmental psychology and environmental planning: collaborative research at the Berkley Environmental Simulation Laboratory', paper presented at Conference on the Federally Coordinated Program of Research and Development in Highway Transportation, US Department of Transportation, Ohio State University, Columbus, OH, November.

Crampton, J.W. 2001. 'Maps as social constructions: power, communication and visualization', *Progress in Human Geography*, 25(2): 235–52.

Crary, J. 1995. *Techniques of the Observer*, Cambridge, MA: MIT Press.

Crary, J. 1999. *Suspensions of Perception: Attention, Spectacle and Modern Culture*, Cambridge, MA and London: MIT Press.

Crary, J. and Kwinter, S. (eds) 1992. *Incorporations*, Zone 6, Cambridge, MA: MIT Press.

Cravey, A.J., Arcury, T.A. and Quandt, S.A. 2000. 'Mapping as a means of farmworker education and empowerment', *Journal of Geography*, 99: 229–37.

Curry, M. 1991. 'On the possibility of ethics in geography: writing, citing and the constitution of intellectual property', *Progress in Human Geography*, 15: 125–47.

Curry, M. 1996. *The Work in the World: Geographical Practice and the Written Word*, Minneapolis: University of Minnesota Press.

Curry, M. 1998. *Digital Places: Living with Digital Information Technologies*, London: Routledge.

Daguerre, L.J.M. 1968. *The History of the Diorama and the Daguerreotype*, New York: Dover.

Dangermond, J. 1991. 'Business adapting GIS to a host of applications', *GIS World*, October, 51–6.

de Seta, C. 1994a. 'The tale of two cities: Siena and Venice', in H.A. Millon and V. Magnago Lampugnani (eds), *The Renaissance from Brunelleschi to Michelangelo: The Representation of Architecture*, London: Thames & Hudson.

de Seta, C. 1994b. 'Parthenopean panopticon: Edgar George Papworth', *Magazine of Franco Maria Ricci*, 68 (June): 97–126.

de Souza Santos, B. 1991. 'Una cartografia simbolica de las representaciones sociales', *Nueva Sociedad*, 116 (Caracas), November–December: 23.

Deleuze, G. 1988. *Foucault*, Minneapolis and London: University of Minnesota Press.

Deleuze, G. and Guattari, F. 1983. *Anti-Oedipus: Capitalism and Schizophrenia*, Minneapolis: University of Minnesota Press.

Deleuze, G. and Guattari, F. 1987. *A Thousand Plateaus: Capitalism and Schizophrenia*, Minneapolis: University of Minnesota Press.

Derrida, J. 1976. *Of Grammatology*, trans. G.C. Spivak, Baltimore, MD: The Johns Hopkins University Press.

Derrida, J. 1981. *Positions*, trans. A. Bass, London: Athlone.

Derrida, J. 1986. 'But, beyond…', in H.L. Gates (ed.), *'Race', Writing and Difference*, Chicago and London: University of Chicago Press: 354–69.

Derrida, J. 1991. 'Speech and phenomena', in P. Kamuf (ed.), *A Derrida Reader: Between the Lines*, New York: Columbia University Press.

Derrida, J. 1994. *Specters of Marx: The State of the Debt, the Work of Mourning and the New International*, trans. Peggy Kamuf, London: Routledge.

Derrida, J. 1998. *Monolingualism of the Other or the Prosthesis of Origin*, trans. P. Mensah, Stanford: Stanford University Press.

Descargues, P. 1997. *Perspective*, New York: Harry N. Abrams Publishers.

Descartes, R. 1965. *Discourse on Method, Optics, Geometry and Meteorology*, Indianapolis: Hackett Publishing.

Dobson, J. 1983. 'Automated geography', *The Professional Geographer*, 35(2): 135–43.

Dobson, J. 1993. 'The geographic revolution: a retrospective on the age of automated geography', *Professional Geographer*, 45(4): 431–9.

Dodd, A. 2000. '"Unacceptable renewals": the geopolitics of Martian cartography', *M/C: A Journal of Media and Culture*. 3, 6 http://moby.curtin.edu.au/~ausstud/mc/0012/mars.html

Dodge, M. n.d. *An Atlas of Cyberspaces*, www.cybergeography.org/atlas/atlas.html

Doel, M. 1999. *Poststructuralist Geographies: The Diabolical Art of Spatial Science*, Lanham, MD: Rowman and Littlefield.

Dorn, M.L. 2002. 'Climate, alcohol and the American body politic: the medical and moral geographies of Daniel Drake (1785–1852)', Ph.D. dissertation, Department of Geography, University of Kentucky, Lexington, KY.

Downs, R.M. and Stea, D. (eds) 1973. *Image and Environment: Cognitive Mapping and Spatial Behaviour* Chicago: Aldine.

Downs, R.M. and Stea, D. 1977. *Maps in Minds: Reflections on Cognitive Mapping*, New York: Harper & Row.

Dreyfus, H.L. 1992. *What Computers Still Can't Do: A Critique of Artificial Reason*, Cambridge, MA: MIT Press.

Driver, F. 2001. *Geography Militant: Cultures of Exploration and Empire*, London: Blackwell Publishers.

Edgerton, Jr., S.Y. 1975. *The Renaissance Rediscovery of Linear Perspective*, New York: Basic Books.

Editorial comment. 1992. 'Politics in maps, maps in politics: a tribute to Brian Harley', *Political Geography*, 11(2), March: xx–xxx.

Edney, M.H. 1989. 'Reply to "Deconstructing the map"', *Cartographica*, 26(3/4): 93–5.

Edney, M.H. 1997. *Mapping an Empire: The Geographical Construction of British India 1765–1943*, Chicago: Chicago University Press.

Edney, M.H. 1999. 'Reconsidering Enlightenment geography and map-making: Reconnaissance, mapping, archive', in C.W.J. Withers and D.N. Livingstone (eds), *Geography and Enlightenment*, Chicago: Chicago University Press.

Ellul, J. 1967. *The Technological Society*, New York: Random House.

Espenshade, Jr., E.B. 1959. 'Cartographic developments and new maps', in P.E. James (ed.), *New Viewpoints in Geography*, Washington, DC: National Council for the Social Studies, Twenty-Ninth Yearbook: 93–4.

Evans, D.S. and Leder, M.R. 1999. 'Economics for the third industrial revolution', *Viewpoint*, 1, http://www.mmc.com/views/99sum.evans.shtml

Farinelli, F. 1998. 'Did Anaximander ever say (or write) any words? The nature of cartographical reason', *Ethics, Place and Environment*, 1(2): 135–44.

Farinelli, F. 1999. 'Text and image in 18th and 19th-century German geography: the *Witz* of landscape and the astuteness of representation', in A. Buttimer, S.D. Brunn and U. Wardenga (eds), *Text and Image: Social Construction of Regional Knowledges*, Leipzig: Institut für Länderkunde: 38–45.

Feldman, A. 1993. *Formations of Violence. The Narrative of the Body and Political Terror in Northern Ireland*, Chicago and London: University of Chicago Press.

Figlio, K. 1996. 'Knowing, loving and hating nature: a psychoanalytic view', in G. Robertson *et al.* (eds), *Future Nature: Nature/Science/Culture*, London and New York: Routledge.

Fontaine, G. 1992. 'The experience of a sense of presence in intercultural and international encounters', *Presence*, 1(4): 482–90.

Foucault, M. 1973. *The Order of Things: An Archaeology of the Human Sciences*, New York: Vintage Books.

Foucault, M. 1977/1979. *Discipline and Punish. The Birth of the Prison*, trans. A. Sheridan, New York: Pantheon.

Foucault, M. 1980. 'Two lectures', *Power/Knowledge: Selected Interviews and Other Writings 1972–1977*, New York: Pantheon: 78–108.

Foucault, M. 1984. *The History of Sexuality Volume 1*, trans. Robert Hurley, New York: Vintage Books.

Foucault, M. 1986. 'Of other spaces', *Diacritics*, 16(1), spring: 22–7.

Foucault, M. 1991. 'Discourse on power', *Remarks on Marx. Conversations with Duccio Trombadori*, trans. R. James Goldstein and James Cascaito, New York: Semiotexte: 173–4.

Foucault, M. 2000. 'The subject and power', in J.D. Faubion (ed.), *Power: Essential Works of Michel Foucault, Vol. 3 1954–84*, trans. R. Hurley and others, New York: New Press.

Francica, J.R. 1991. 'GIS in business', *GIS World*, October: 49.

Fraser, N. 1989. *Unruly Practices: Power, Discourse and Gender in Contemporary Social Theory*, Minneapolis: University of Minnesota Press.

Fukuyama, F. 1992. *The End of History and the Last Man*, New York: Avon Books.

Fyfe, G. and Law, J. 1988. *Picturing Power: Visual Depiction and Social Relations*, London and New York: Routledge.

Gadamer. H.G. 1981. *Reason in the Age of Science*, Cambridge, MA: MIT Press.

Gates, B. 1996. *The Road Ahead*, New York: Penguin.

Gatrell, A. 1983. *Distance and Space: A Geographical Perspective*, Oxford: Clarendon Press.

Gelernter, D. 1992. *Mirror Worlds or the Day Software Puts the University in a Shoebox... How It Will Happen and What It Will Mean*, New York and Oxford: Oxford University Press.

Genette, G. 1982. 'Stendhal' (1968), in A. Sheridan (ed.), *Figures of Literary Discourse*, New York: Columbia University Press.

Gibson-Graham, J.K. 1996. *The End of Capitalism (as we know it): A Feminist Critique of Political Economy*, London: Blackwell.

Giddens, A. 1990. *The Consequences of Modernity*, Stanford: Stanford University Press.

Gillispie, C.C. 1980. *Science and Polity in France at the End of the Old Regime*, Princeton, NJ: Princeton University Press.

Godlewska, A. 1995. 'Map, text, and image: the mentality of Enlightenment conquerors: a new look at the *Description de l'Egypte*', *Transactions of the Institute of British Geographers, NS*, 20: 5–28.

Godlewska, A.M. 1999a. *Geography Unbound: French Geographic Science from Cassini to Humboldt*, Chicago: University of Chicago Press.

Godlewska, A.M.C. 1999b. 'From Enlightenment vision to modern science? Humboldt's visual thinking', in D.N. Livingstone and C.W.J. Withers (eds), *Geography and Enlightenment*, Chicago and London: University of Chicago Press, 236–79.

Goodchild, M.F. 2000. 'Cartographic futures on a digital earth', *Cartographic Perspectives*, 36: 3–11.

Goodman, N. 1978. *Ways of Worldmaking*, Indianapolis and Cambridge, MA: Hackett Publishing Co.

Goss, J. 1995. 'Marketing the new marketing: the strategic discourse of geodemographic information systems', in J. Pickles (ed.), *Ground Truth: The Social Implications of Geographical Information Systems*, New York: Guilford Press: 130–70.

Gould, P. 1969. 'The new geography', *Harper's Magazine*, March: 91–100.

Gould, P. 1982. 'Guest editorial: things I do not understand very well: I. Mathematics and thinking in the human sciences', *Environment and Planning A*, 14: 1279–82.

Gould, P. 1985. *The Geographer at Work*, London: Routledge and Kegan Paul.

Gould, P. and Pitts, F. 2002. *Geographical Voices*, Syracuse: Syracuse University Press.

Gould, P. and White, R. 1986. *Mental Maps*, Boston: Allen & Unwin.

Gramsci, A. 1981. *Selections from the Prison Notebooks*, trans. G.N. Smith and Q. Hoare, New York: International Publishers.

Gregory, D. 1994. *Geographical Imaginations*, Oxford: Blackwell.

Guattari, F. 1991. 'Regimes, pathways, subjects', in J. Crary and S. Kwinter (eds), *Incorporations*, Zone 6, New York: Urzone.

Guelke, L., (ed.) 1977. 'The nature of cartographic communication', Monograph 19, *Cartographica*.

Guelke, L. 1981. 'Maps in modern geography: geographical perspectives on the new cartography', Monograph 27, *Cartographica*, 18(2).

Gusso, J.S. and Lasala, V. 1991. 'Desktop mapping delivers GIS efficiency to telecommunication', *GIS World*, October: 86–8.

Habermas, J. 1968/1971. *Knowledge and Human Interests*, Boston: Beacon Press.

Habermas, J. 1973. *Theory and Practice*, Boston: Beacon Press.

Habermas, J. 1994. *The Past as Future*, Lincoln: University of Nebraska Press.

Habermas, J. 1997. *A Berlin Republic: Writings on Germany*, Lincoln: University of Nebraska Press.

Hacking, I. 1982. 'Language, truth, and reason', in M. Hollis and S. Lukes (eds), *Rationality and Relativism*, Cambridge, MA: MIT Press: 48–66.

Hacking, I. 1992a. 'Statistical language, statistical truth, and statistical reason: the self-authentification of a style of scientific reasoning', in E. McMullin (ed.), *The Social Dimensions of Science*, Notre Dame, IN: University of Notre Dame Press: 130–57.

Hacking, I. 1992b. ' "Style" for Historians and Philosophers', *Studies in History and Philosophy of Science*, 23(1): 1–20.

Hall, S.S. 1993. *Mapping the Next Millennium: How Computer-Driven Cartography is Revolutionizing the Face of Science*, New York: Vintage Books.

Hanke, R. 1992. 'The first casualty?' *Public 6:* Violence: 134–40.

Hannah, M.G. 2000. *Governmentality and the Mastery of Territory in Nineteenth-Century America*, Cambridge, MA: Cambridge University Press.

Haraway, D. 1989. 'Gender, race and nature in the world of modern science', *Primate Visions*, New York: Routledge.

Haraway, D. 1991. 'A cyborg manifesto: science, technology and socialist-feminism in the late twentieth century', *Simians, Cyborgs and Women: The Reinvention of Nature*, London: Routledge.

Harbison, R. 1977. *Eccentric Spaces*, New York: Random House.

Harley, B. 1986. '*Imago Mundi*/the first fifty years and the next ten', *Cartographica*, 23(3), autumn: 1–15.

Harley, B. 1987. 'The map and the development of the history of cartography', in B. Harley and D. Woodward (eds), *The History of Cartography Vol. 1: Cartography in Prehistoric, Ancient and Medieval Europe and the Mediterranean*, Chicago and London: University of Chicago Press: 1–42.

Harley, B. 1988a. 'Silences and secrecy: the hidden agenda of cartography in early modern Europe', *Imago Mundi*, 40: 57–76.

Harley, B. 1988b. 'Maps, knowledge and power', in D. Cosgrove and S.J. Daniels (eds), *The Iconography of Landscape*, Cambridge Studies in Historical Geography, 9, Cambridge, MA: Cambridge University Press: 277–312.

Harley, J.B. 1989a. 'Historical geography and the cartographic illusion', *Journal of Historical Geography*, 15: 80–91.

Harley, J.B. 1989b. 'Deconstructing the map', *Cartographica*, 26(2): 1–20.

Harley, J.B. 1990. 'Cartography, ethics and social theory', *Cartographica*, 27(2): 1–23.

Harley, J.B. 1992a. 'Deconstructing the map', in T. Barnes and J. Duncan (eds), *Writing Worlds: Discourse, Text and Metaphor in the Representation of Landscape*, London: Routledge: 231–47.

Harley, J.B. 1992b. 'Rereading the maps of the Columbian encounter', *Annals of the Association of American Geographers*, 82(3): 522–42.

Harley, J.B. 2001. *The New Nature of Maps: Essays in the History of Cartography*, Paul Laxton (ed.), Baltimore and London: Johns Hopkins University Press.

Harley, J.B. and Laxton, P. (eds) 2002. *The New Nature of Maps: Essays in the History of Cartography*, Baltimore: Johns Hopkins University Press.

Harley, J.B. and Woodward, D. (eds) 1992. *The History of Cartography*, Chicago: The University of Chicago Press.

Hartshorne, R. 1939. *The Nature of Geography: A Critical Survey of Current Thought in the Light of the Past*, Lancaster, PA: Association of American Geographers.

Harvey, D. 1969. *Explanation in Geography*, London: Edward Arnold.

Harvey, D. 1974. 'What kind of geography for what kind of public policy?' *Transactions of the Institute of British Geographers*, 63: 18–24.

Harvey, D. 1989. *The Condition of Postmodernity: An Inquiry Into the Origins of Cultural Change*, Oxford: Basil Blackwell.

Harvey, D. 2000. 'Cartographical identities: geographical knowledges under globalization', *Spaces of Capital*, London: Routledge: 208–33.

Harvey, P.D.A. 1993. *Maps in Tudor England*, Chicago: University of Chicago Press.

Haushofer, K. 1928. 'Die suggestive Karte', *Bausteine zur Geopolitik*, Berlin: Kurt Vowinckel Verlag: 343–8.

Heidegger, M. 1928. *Being and Time*, New York: Harper & Row.

Heidegger, M. 1969. *Identity and Difference*, trans. J. Stambaugh, New York: Harper & Row.

Heidegger, M. 1977. 'The question concerning technology', in *The Question Concerning Technology and Other Essays*, New York: Harper Colophon: 3–35.

Heidegger, M. 1982. *The Question Concerning Technology and Other Essays*, New York: HarperCollins.

Heim, M. 1992. 'The erotic ontology of cyberspace', in M. Benedikt (ed.), *Cyberspace: First Steps*, Boston: MIT Press: 59–80.

Heisenberg, W. 1959. *Physics and Philosophy: The Revolution in Modern Science*, London: George Allen & Unwin (World Perspectives 19).

Heivly, C. 1991. 'Route planning tool boosts saturation delivery efficiency.', *GIS World*, October: 78–9.

Henricksen, A.K. 1994. 'The power and politics of maps', in G.J. Demko and W.B. Wood (eds), *Reordering the World: Geopolitical Perspectives on the Twenty-first Century*, Boulder: Westview Press: 49–70.

Herb, G.H. n.d. 'Before the Nazis: maps as weapons in German nationalist propaganda', *Mercator's World* www.mercatormag.com/article.php3?I=60

Herb, G.H. 1989. 'Persuasive cartography in *Geopolitik* and national socialism', *Political Geography Quarterly*, 8(3) July: 289–303.

Herb, G.H. 1997. *Under the Map of German Nationalism and Propaganda 1918–1945*, London: Routledge.

Hillis, K. 1999a. 'Toward the light within: spatial metaphors, new optical media and changing subjectivities', in M. Crang, P. Crang and J. May (eds), *Virtual Geographies: Bodies, Space and Relations*, London: Routledge: 23–43.

Hillis, K. 1999b. *Digital Sensations: Space, Identity and Embodiment in Virtual Reality*, Minneapolis: University of Minnesota Press.

Hinsley, C.M. 1991. 'The world as marketplace: commodification of the exotic at the World's Columbian Exposition, Chicago, 1893', in I. Karp and S.D. Lavine (eds), *Exhibiting Cultures: The Poetics and Politics of Museum Display*, Washington and London: Smithsonian Institution Press: 344–65.

Hitt, J. 1995. 'Atlas shrugged: the new face of maps', *Lingua Franca*, 5(5) July/August: 24–33.

Holmes, O.W. 1859. 'The stereoscope and the stereograph', *Atlantic Monthly*, 3 (June): 738–48, www.yale.edu/amstud/inforev/stereo.html

Horkheimer, M. and Adorno, T.W. 1944. *The Dialectic of Enlightenment*, trans. J. Cumming, New York: Continuum (1991).

Huxhold, W.E. 1991. 'The GIS profession: title, pay, qualifications', *Geo Info Systems*, March, 12: 22.

James, P.E. Quoted in Hartshorne, R. 1939. *The Nature of Geography*, Lancaster, PA: Association of American Geographers, p. 248.

Jameson, F. 1971. *Marxism and Form: Twentieth Century Dialectical Theories of Literature*, Princeton, NJ: Princeton University Press.

Jameson, F. 1984. 'Postmodernism, or the cultural logic of late capitalism', *New Left Review*, 146: 53–92.

Jameson, F. 1992. *Signatures of the Visible*, New York and London: Routledge.

Jarvis, B. 1998. *Postmodern Cartographies: The Geographical Imagination in Contemporary American Culture*, New York: St Martin's.

Jay, M. 1994. *Downcast Eyes: The Denigration of Vision in Twentieth-Century French Thought*, Los Angeles: University of California Press.

Johnson, H.B. 1976. *Order upon the Land: The US Rectangular Land Survey in the Upper Mississippi Country*, New York: Oxford University Press.

Jowett, R., Kain, J.P. and Baigent, E. 1992. *The Cadastral Map in the Service of the State: A History of Property Mapping*, Chicago and London: The University of Chicago Press.

Keating, R. 1992. 'Building accuracy into GIS', *GIS World*, 5(5): 32–4.

Kellner, D. 1989a. 'Media, simulations and the end of the social', *Jean Baudrillard: From Marxism to Postmodernism and Beyond*, Stanford: Stanford University Press: 60–92.

Kellner, D. 1989b. *Critical Theory, Marxism and Modernity*, Baltimore: The Johns Hopkins Press.

Kellner, D. 1990. *Television and the Crisis of Democracy*, Boulder: Westview Press.

Kimerling, A.J. 1989. 'Cartography', in G.L. Gaile and C.J. Willmott (eds), *Geography in America*, Columbus: Merrill: 686–718.

King, G. 1996. *Mapping Reality: An Exploration of Cultural Cartographies*, London: Macmillan.

Kirchner, R. and Thomas, R.K. 1991. 'Dunkin' Donuts plugs hole in location strategy', *GIS World*, October: 89.

Kirsch, S. 1999. 'Science out of bounds? Regional surveys and cartographies of power in the American West', paper presented at the Western Humanities Conference, University of California, San Diego, 14–16 October http://wha.ucdavis.edu/1999/Scott-Kirsch.htm

Kockelmans, J.J. 1967. *Phenomenology: The Philosophy of Edmund Husserl and Its Interpretation*, New York: Doubleday.

Krishna, S. 1996. 'Cartographic anxiety: mapping the body politic in India', in M.J. Shapiro and H.R. Alker (eds), *Challenging Boundaries: Global Flows, Territorial Identities*, Borderlines Series, vol. 2, Minneapolis: University of Minnesota Press: 193–214.

Kroker, A. 1992. *The Possessed Individual: Technology and the French Postmodern*, New York: St Martin's Press.

Krygier, J.B. 1997. 'Envisioning the American West: maps, the representational barrage of the 19th century expedition reports, and the production of scientific knowledge', *Cartography and Geographic Information Systems*, 24(1): 27–50.

Lake, R.W. 1993. 'Planning and applied geography: positivism, ethics and geographic information systems', *Progress in Human Geography*, 17(3): 404–13.

Landow, G.P. 1992. *Hypertext: The Convergence of Contemporary Critical Theory and Technology*, Baltimore: The Johns Hopkins Press.

Lasswell, H.D. 1927. *Propaganda Techniques in World War I*, Cambridge, MA: MIT Press (1971).

Latour, B. 1988. *The Pasteurization of France*, trans. Alan Sheridan and John Law, Cambridge, MA: Harvard University Press.

Latour, B. 1993. *We Have Never Been Modern*, Cambridge, MA: MIT Press.

Latour, B. 1999. *Pandora's Hope: Essays on the Reality of Science Studies*, Cambridge, MA and London: Harvard University Press.

Lefebvre, H. 1991. *The Production of Space*, trans. D. Nicholson-Smith, Oxford: Blackwell.

Lestringant, F. 1994. *Mapping the Renaissance World: The Geographical Imagination in the Age of Discovery*, trans. David Fausett, Berkeley: University of California Press.

Lewis, G.H. 1987. 'Misinterpretation as a source of error on Euro-American maps', *Annals of the Association of American Geographers*, 77: 542–63.

Lewis, P. 1985. 'Presidential address: beyond description', *Annals of the Association of American Geographers*, 75(4): 465–77.

Lipietz, A. 1992. *Towards a New Economic Policy*, Oxford: Oxford University Press.

Livingstone, D.N. 1992. *The Geographical Tradition: Episodes in the History of a Contested Tradition*, Oxford: Blackwell.

Luke, T.W. 1993. 'Beyond Leviathan, Beneath Lilliput: geopolitics and globalization', presented at the Annual Meeting of the Association of American Geographers, Atlanta, 7–9 April.

Luke, T.W. and White, S.K. 1985. 'Critical theory, the information revolution and an ecological path to modernity', in J.M. Forester (ed.), *Critical Theory and Public Life*, Cambridge, MA: MIT Press: 22–53.

Lynch, K. *The Image of the City*, Cambridge, MA: MIT Press.

Lyon, D. 1994. *The Electronic Eye: The Rise of Surveillance Society*, Minneapolis: University of Minnesota Press.

Lyotard, J.F. 1984. *The Postmodern Condition: A Report on Knowledge*, trans. G. Bennington and B. Massumi, Minneapolis: University of Minnesota.

MacEachren, A.M. 1994. *Some Truth with Maps: A Primer on Symbolization and Design*, Washington: Association of American Geographers.

MacEachren, A.M. 1995. *How Maps Work: Representation, Visualization and Design*, New York: Guilford Press.

Maffini, G. 1991. 'GIS at threshold of business applications', *GIS World*, October: 50–2.

Maguire, D.J., Goodchild, M.F. and Rhind, D.W. 1991. *Geographical Information Systems*, London: Longman Scientific and Technical, New York: John Wiley & Sons, 2 vols.

Mahon, R. 1987. '"From Fordism to?" New technology, labour markets and unions', *Economic and Industrial Democracy*, 8: 5–60.

Mann, M. 1986. *The Sources of Social Power*, vol. 1, New York: Cambridge University Press.

Marino, J. 1992. 'Administrative mapping in the Italian states', in D. Buisseret (ed.), *Monarchs, Ministers and Maps*, Chicago: University of Chicago Press: 5–25.

Mark, D.N. 1997. 'The history of GIS: invention and re-invention of triangulated irregular networks (TINS)', *Proceedings GIS/LIS*, www.geog.buffalo.edu/ncgia/gishist/GISLIS97.html

Mark, D.N., Chrisman, N., Frank, A.U., McHaffie, P.H. and Pickles, J. 'The GIS History Project', unpublished mimeo: http://www.geog.buffalo.edu/ncgia/gishist/bar.harbor.html

Martin, D. 1991. *Geographic Information Systems and Their Socioeconomic Applications*, London: Routledge.

Martin, E.L. and Rinalducci, E.J. 1983. 'Low-level flight simulation: vertical cues (AFHRL-TR-83-17, AD-A133612)', Williams Air Force Base, AZ: Operations Training Division, Human Resources Laboratory.

Massey, D. 1994. *Space, Place and Gender*, Minneapolis: University of Minnesota Press.

Matless, D. 1999. 'The uses of cartographic literacy', in D. Cosgrove (ed.), *Mappings*. London: Reaktion Books: 198–212.

Mattelart, A. 1999. 'Mapping modernity: utopia and communications networks', in D. Cosgrove (ed.), *Mappings*, London: Reaktion Books: 169–92.

McDermott, P.D. 1969. 'Cartography in advertising', *Canadian Cartographer*, 6(2) December: 149–55.

McHaffie, P. 1995. *Ground Truth: The Social Implications of Geographical Information Systems*, New York: Guilford.

McHaffie, P. 1997. 'Decoding the globe: globalism, advertising and corporate practice', *Environment and Planning D: Society and Space*, 15: 73–86.

Merleau-Ponty, M. 1964. 'Cézanne's doubt', *Sense and Non-sense*, Evanston, IL: Northwestern University Press.

Mikhova, D. and Pickles, J. 1994. 'GIS in Bulgaria: prospects and problems', *International Journal of Geographical Information Systems*, 8(5) May: 471–7.

Mills, C.W. 1956. *The Power Elite*, New York: Oxford University Press.

Miller, A. 1996. 'The panorama, the cinema, and the emergence of the spectacular', *Wide Angle*, 18(2), April: 34–69.

Mitchell, T. 1991. *Colonizing Egypt*, Los Angeles: University of California Press.

Moloney, T. and Dellavedova, B. 1991. 'Retail applications reap the benefits of GIS', *GIS World*, October: 62–6.

Moncla, A. and McConnell, L. 1991. 'TIGER helps retailers address marketing issues', *GIS World*, October: 80–2.

Monmonier, M. 1975. 'Maps, distortion and meaning', Association of American Geographers Resource Paper No. 75-4, Washington, DC.

Monmonier, M. 1985. *Technological Transition in Cartography*, Madison, WI: University of Wisconsin Press.

Monmonier, M. 1989. *Maps with the News: The Development of American Journalistic Cartography*, Chicago: University of Chicago Press.

Monmonier, M. 1991. *How to Lie with Maps*, Chicago: University of Chicago Press.

Monmonier, M. 1993. *Mapping It Out: Expository Cartography for the Humanities and Social Sciences*, Chicago: University of Chicago Press.

Monmonier, M. 1995. *Drawing the Line: Tales of Maps and Cartocontroversy*, New York: Henry Holt and Co.

Monmonier, M. 1997. *Cartographies of Danger: Mapping Hazards in America*, Chicago: University of Chicago Press.

Monmonier, M. 1999. *Air Apparent: How Meteorologists Learned to Map, Predict and Dramatize Weather*, Chicago: University of Chicago Press.

Monmonier, M. 2000. 'History of mapping and map use in the twentieth century: an invitation', *Cartographic Perspectives*, 35 (winter): 3–6.

Monmonier, M. 2001. *Bushmanders and Bullwinkles: How Politicians Manipulate Electronic Maps and Census Data to Win Elections*, Chicago: University of Chicago Press.

Monmonier, M. 2002. 'Maps, politics, and history', Jeremy Crampton talks with Mark Monmonier, by Jeremy Crampton, *Environment and Planning D: Society and Space*, 20(6), December: 637–46.

Monmonier, M. and Schnell, G.A. 1988. *Map Appreciation*, Englewood Cliffs, NJ: Prentice-Hall.

Morton, O. 1999. 'A launch for the little guy: satellite technology can now help real people', *Newsweek*, 4 October.

Muehrcke, P. 1972. *Thematic Cartography*, Washington: Association of American Geographers.

Mundy, B.E. 1996. *The Mapping of New Spain: Indigenous Cartography and the Maps of the Relaciones Geograficas*, Chicago: University of Chicago Press.

Murdock, G. and Golding, P. 1989. 'Information poverty and political inequality: citizenship in the age of privatized communications', *Journal of Communication*, 39: 180–96.

Musée des Beaux-Arts Lille. 1989. *Plans en Relief: Villes Fortes des Anciens Pays-Bas Français au XVIIIeS*, Musée des Beaux-Arts Lille.

Nelson, T.H. 1992. 'Virtual world without end: the story of Xanadu', in L. Jacobson (ed.), *Cyberarts: Exploring Art and Technology*, San Francisco: Miller Freeman: 157–69.

Newell, R.G. and Theriault, D.G. 1990. 'Is GIS just a combination of CAD and DBMS?', *Mapping Awareness*, 4 (3): 42–5.

Norris, C. 1987. *Derrida*, Cambridge, MA: Harvard University Press.

Olsson, G. 1972. 'Some notes on geography and social engineering', *Antipode*, 4(1): 1–21.

Olsson, G. 1974. 'Servitude and inequality in spatial planning: ideology and methodology in conflict', *Antipode*, 6(1), 16–21.

Olsson, G. 1991. 'Invisible maps: a prospective', *Geografiska Annaler*, 73(B)(1): 85–92.

Olsson, G. 1992a. 'Lines of power', in T.J. Barnes and J.S. Duncan (eds), *Writing Worlds: Discourse, Text and Metaphor in the Representation of Landscape*, London: Routledge: 86–96.

Olsson, G. 1992b. *Lines of Power/Limits of Language*, Minneapolis: University of Minnesota Press.

Olsson, G. 1994. 'Heretic cartography', *Fennia*, 172(2): 115–30.

Olsson, G. 1998. 'Towards a critique of cartographic reason', *Ethics, Place and Environment*, 1(2): 145–55.

Olsson, G. 1999. 'Washed in a washing machine™', in C. Minca (ed.), *Postmodern Geographical Praxis*, Oxford: Blackwell.

Olsson, G. 2000. 'From a = b to a = a', *Environment and Planning A*, 37(7) July: 1235–44.

Olsson, G. 2002. 'Glimpses', in P.R. Gould and F. Pitts (eds), *Geographical Voices*, Syracuse: Syracuse University Press.

Openshaw, S. 1991. 'Commentary: a view on the GIS crisis in geography, or, using GIS to put Humpty-Dumpty back together again', *Environment and Planning A*, 23: 621–8.

Openshaw, S. 1992. 'Commentary: further thoughts on geography and GIS: a reply', *Environment and Planning A*, 24: 463–6.

Orlove, B. 2002. *Lines in the Water: Nature and Culture in Lake Titicaca*, Berkeley and Los Angeles: University of California Press.

O Tuathail, G. 1993a. 'The new East–West conflict? Japan and the Bush administration's "New World Order"', *Area*, 25(2): 127–35.

O Tuathail, G. 1993b. 'The effacement of place? US foreign policy and the spatiality of the Gulf crisis', *Antipode*, 25(1): 4–31.

Ould-Mey, M. 1994. 'Global adjustment: implications for peripheral states', *Third World Quarterly*, 15(2): 319–36.

Pagels, H.R. 1989. *The Dreams of Reason: The Computer and the Rise of the Sciences of Complexity*, New York: Bantam.

Paulston, R.G. 1996. *Social Cartography: Mapping Ways of Seeing Social and Environmental Change*, New York and London: Garland Publishing.

Paulston, R.G. 1997. 'Mapping visual culture in comparative education discourse', *Compare*, 27(2): 117–52.

Pavlinek, P. and Pickles, J. 2000. *Environmental Transitions: Transformation and*

Ecological Defense in Central and Eastern Europe, London and New York: Routledge.

Payne, A. 1992. 'War in the age of intelligent machines. An interview with Manuel Delanda', *Public 6*, 'Violence': 127–34.

Peet, R. 1993a. 'Reading Fukuyama: politics at the end of history', *Political Geography*, 12(1), January: 64–78.

Peet, R. 1993b. 'The end of prehistory and the first man', *Political Geography*, 12(1) January: 91–5.

Peet, R. and Hartwick, E. 1999. *Theories of Development*, New York: The Guilford Press.

Peet, R. and Hartwick, E. 2002. 'Poststructural thought policing', *Economic Geography*, 78(1): 87–8.

Penley, C. and Ross, A. 1990. 'Cyborgs at large: interview with Donna Haraway', *Social Text*, 25–6.

Penley, C. and Ross, A. (eds) 1991. *Technoculture*, Minneapolis: University of Minnesota Press.

Pequet, D.J. 2002. *Representations of Space and Time*, New York and London: The Guilford Press.

Petchenick, B.B. 1983. 'A map maker's perspective on map design research', in D.R.F. Taylor (ed.), *Graphic Communication and Design in Contemporary Cartography*, Chichester: John Wiley & Sons.

Phillips, R. 1997. *Mapping Men and Empire: A Geography of Adventure*, London: Routledge.

Pickering, A. 1995. *The Mangle of Practice: Time, Agency and Science*, Chicago: University of Chicago Press.

Pickles, J. 1985. *Phenomenology, Science and Geography: Space and the Human Sciences*, Cambridge, MA: Cambridge University Press.

Pickles, J. 1986a. *Geography and Humanism*, Sidcup, Kent: CATMOG/Elsevier.

Pickles, J. 1986b. 'Geographic theory and education for democracy', *Antipode*, 18(2), August: 136–54.

Pickles, J. 1991. 'Geography, GIS, and the surveillant society', in J.W. Frazier, B.J. Epstein, F.A. Schoolmaster III and H.E. Moon (eds), *Papers and Proceedings of the Applied Geography Conferences*, vol. 14: 80–91.

Pickles, J. 1992a. 'Review of *Geographic Information Systems and Their Socioeconomic Applications*', *Environment and Planning D: Society and Space*, 10: 597–606.

Pickles, J. 1992b. 'Texts, hermeneutics and propaganda maps', in T.J. Barnes and J.S. Duncan (eds), *Writing Worlds: Discourse, Text and Metaphor in the Representation of Landscape*, London and New York: Routledge: 193–230.

Pickles, J. 1993. 'Discourse on method and the history of discipline: reflections on Jerome Dobson's 1993 *Automated Geography*', *Professional Geographer*, 45(4): 451–5.

Pickles, J. 1994. 'Cyber-empires: cultural politics and democratic theory in the age of electronic representation and the virtual sign', Work-in-Progress Series, Committee on Social Theory, University of Kentucky.

Pickles, J. 1995. *Ground Truth: The Social Implication of Geographic Information Systems*, New York: The Guilford Press.

Pickles, J. 1997. 'Tool or science? GIS, technoscience, and the theoretical turn', *The Annals of the Association of American Geographers*, 87(2) June: 363–72.

Pickles, J. 1998a. 'The production of new identities: representing space, nature, and subjects', keynote presentation to the Annual Conference of NACIS, Lexington KY.

Pickles, J. 1998b. 'Representing subjects', keynote presentation to the Colloque: Dire, Ecrire et Figurer L'Espace, 4–5 December, Université François-Rabelais, Tours.

Pickles, J. 1999. 'Cartography, digital transitions and questions of history', *Proceedings of the 19th International Cartographic Conference.* Ottawa.

Pickles, J. 2000a. 'Arguments, debates and dialogues: The GIS–social theory debate and the concern for alternatives', in P. Longley, M. Goodchild, D. Maguire and D. Rhind (eds), *Geographical Information Systems: Principles, Techniques, Management, and Applications,* New York: John Wiley: 49–60.

Pickles, J. 2000b. 'Cartography, digital transitions and questions of history', *Cartographic Perspectives,* 37 (autumn): 4–18.

Pickles, J. 2000c. 'Social and cultural cartographies and the spatial turn in social theory', *Journal of Historical Geography,* 25(1), 93–8.

Pickles, J. 2001. (Book review essay) *'Development "Deferred": Poststructuralism, Postdevelopment and the Defense of Critical Modernism',* Economic Geography, October 77(4): 383–8.

Pickles, J. 2002. 'Reading development: a response', *Economic Geography,* January 78(1): 89–90.

Pickles, J. and Mikhova, D. 1998. 'The political economy of environmental data in the Bulgarian transition', in K. Paskaleva, P. Shapira, J. Pickles and B. Koulov (eds), *Bulgaria in Transition: The Environmental Consequences of Political and Economic Transformation,* Aldershot, UK: Ashgate.

Pickles, J. and Smith, A. (eds) 1998. *Theorizing Transition: The Political Economy of Post-Communist Transformations,* London and New York: Routledge.

Pickles, J. and Watts, M. 1992. 'Paradigms of inquiry?', in R.F. Abler, M.G. Marcus and J.M. Olson (eds) *Geography's Inner Worlds: Pervasive Themes in Contemporary American Geography,* New Brunswick, NJ: Rutgers University Press.

Pile, S. and Rose, G. 1992. 'All or nothing? Politics and critique in the modernism/ postmodernism debate', *Environment and Planning D: Society and Space,* 10: 123–36.

Pile, S. and Thrift, N. (eds). 1995. *Mapping the Subject: Geographies of Cultural Transformation,* London: Routledge.

Popke, E.J. 1999. 'Deconstructing apartheid space: negotiating alterity and history in Durban's Cato Manor', Ph.D. dissertation, Department of Geography, University of Kentucky.

Poster, M. 1990. *The Mode of Information: Poststructuralism and Social Context,* Cambridge, MA: Polity Press.

Pratt, M.L. 1992. *Imperial Eyes: Travel Writing and Transculturation,* London: Routledge.

Pred, A. and Watts, M.J. 1992. *Reworking Modernity: Capitalisms and Symbolic Discontent,* New Brunswick, NJ: Rutgers University Press.

Quam, L.O. 1943. 'The use of maps in propaganda', *Journal of Geography,* 42 (January): 21–32.

Rabinow, P. 1989. *French Modern Norms and Forms of the Social Environment,* Cambridge, MA: MIT Press.

Raulet, G. 1991. 'The new utopia: communication technologies', *Telos*, 87, spring, 39–58.

Reichert, D. (ed.) 1996. *Raumliches Denken*, Hochschulverlag AG an der ETH: Zurich.

Reichert, D. 1998. 'Obituary', *Ethics, Place and Environment*, 1(2): 158–64.

Renner, G.T. 1942. 'Maps for a new world', *Collier's*, 6 June: 14–16, 28.

Renner, G.T. 1944. 'Peace by the map', *Collier's*, 3 June: 44, 47.

Rheingold, H. 1992. *Virtual Reality*, London: Mandarin.

Rhind, D.W. 2000. 'Business, governments, and technology: inter-linked causal factors of change in cartography', *Cartographic Perspectives*, 37 (autumn): 19–25.

Rice, S. 1993. 'Boundless horizons: the panoramic image', *Art in America*, December: 68–71.

Richard, N. 1996. 'The cultural periphery and postmodern decentering: Latin America's reconversion of borders', in J.C. Welchman (ed.), *Rethinking Borders*. Minneapolis: University of Minnesota Press: 71–84.

Ricoeur, P. 1971. 'The model of the text: meaningful action considered as text', *Social Research*, 38(3): 529–62.

Riis, J. 1890. *How the Other Half Lives*, www.yale.edu/amstud/inforev/riis/title.html

Roberts, S. and Schein, R. 1995. 'Earth shattering: global imagery and GIS', in J. Pickles (ed.), *Ground Truth: The Social Implications of Geographical Information Systems*, New York: Guilford: 171–95.

Robinson, A., Sale, R., Morrison, J. and Muehrcke, P.C. 1984. *Elements of Cartography*, New York: Wiley.

Robinson, A.H. 1952. *The Look of Maps: An Examination of Cartographic Design*, Madison: University of Wisconsin Press.

Robinson, A.H. 1982. *Early Thematic Mapping in the History of Cartography*, Chicago: University of Chicago Press.

Robinson, A.H. and Petchenik, B. 1976. *The Nature of Maps*, Chicago: Chicago University Press.

Ronnel, A. 1989. *The Telephone Book: Technology, Schizophrenia, Electric Speech*, Lincoln: University of Nebraska Press.

Rorty, R. 1980. *Philosophy and the Mirror of Nature*, Princeton, NJ: Princeton University Press.

Rose, G. 1993. *Feminism and Geography: The Limits of Geographical Knowledge*, Minneapolis: University of Minnesota Press.

Rose, G. 1995. 'Distance, surface, elsewhere: a feminist critique of the space of phallocentric self/knowledge', *Environment and Planning D: Society and Space*, 13: 761–81.

Rosenthal, P. 1992. 'Remixing memory and desire: the meanings and mythologies of virtual reality', *Socialist Review*, 22(3), July–September: 107–17.

Roszak, T. 1986. *The Cult of Information: The Folklore of Computers and the True Art of Thinking*, New York: Pantheon Books.

Rundstrom, R. 1989. 'A critical appraisal of "applied" geography', in M.S. Kenzer (ed.), *Applied Geography: Issues, Questions and Concerns*, Dordrecht: Kluwer Academic Press: 175–91.

Rundstrom, R. 1990. 'A cultural interpretation of Inuit map accuracy', *Geographical Review*, 80: 156–68.

Rundstrom, R. 1991. 'Mapping postmodernism: indigenous people and the changing direction of North American cartography', *Cartographica*, 28(2), summer: 1–12.

Rundstrom, R. 1993. 'The role of ethics, mapping and the meaning of place in relations between Indians and Whites in the United States', *Cartographica*, 30(1): 21–8.

Runnels, D. 1991. 'Geographic underwriting system streamlines insurance industry', *GIS World*, October: 60–2.

Said, E. 1979. *Orientalism*, New York: Random House.

Salichtchev, K.A. 1978. 'Cartographic communication: its place in the theory of science', *Canadian Cartographer*, 15(2): 93–9.

Schein, R. 1993. 'Representing urban America: 19th-century views of landscape, space, and power', *Environment and Planning D: Society and Space*, 11: 7–21.

Schlichtmann, H. 1985. 'Characteristic traits of the semiotic system "map symbolism"', *The Cartographic Journal*, 22: 23–30.

Schor, N. 1994. 'Collecting Paris', in J. Elsner and R. Cardinal (eds), *The Cultures of Collecting*, Cambridge, MA: Harvard University Press: 252–302.

Scott, J.C. 1998. *Seeing Like a State: How Certain Schemes to Improve the Human Condition Have Failed*, New Haven and London: Yale University Press.

Shapiro, M.J. 1997. *Violent Cartographies: Mapping Cultures of War*, Minneapolis: University of Minnesota Press.

Sheppard, E. 1993. 'Automated geography: what kind of geography for what kind of society', *Professional Geographer*, 45(4): 457–60.

Sismondo, S. 1999. 'Deflationary metaphysics and the natures of maps', paper presented at the UK Conference on Mapping, GIS and Science Studies.

Sismondo, S. and Chrisman, N. 2001. 'Deflationary metaphysics and the natures of maps', *Proceedings of the Philosophy of Science*, 68: 538–49.

Skelton, R.A. 1967. 'Map compilation, production, and research in relation to geographical exploration', in H.R. Friis (ed.), *The Pacific Basin: A History of Its Geographical Exploration*, New York: American Geographical Society.

Smith, J.H. 1989. *Postcard Companion: The Collector's Reference*, Radnor, PA.

Smith, N. 1992. 'Real wars, theory wars', *Progress in Human Geography*, 16(2): 257–71.

Snyder, J.P. 1993. *Flattening the Earth: Two Thousand Years of Map Projections*, Chicago and London: University of Chicago Press.

Soderstrom, O. 1996. 'Paper cities: visual thinking in urban planning', *Ecumene*, 3(3): 249–81.

Soderstrom, O. 1997. 'Spatializing the visual: geographers and the world of images', paper presented at the conference 'Actor–Network Theory and After' 10–11 July 1997, Centre for Social Theory and Technology, Keele University.

Soffner, H. 1942. 'War on the visual front', *The American Scholar*, XI: 465–76.

Soja, E.W. 1989. *Postmodern Geographies: The Reassertion of Space in Critical Social Theory*, London: Verso.

Soja, E.W. and Hooper, B. 1993. 'The spaces that difference makes: some notes on the geographical margins of the new cultural politics', in M. Keith and S. Pile (eds), *Place and the Politics of Identity*, New York: Routledge.

Sonenberg, M. 1989. *Cartographies*, Pittsburgh: University of Pittsburgh Press.

Sparke, M. 1995. 'Between demythologizing and deconstructing the map: Shawnadithit's new-found-land and the alienation of Canada', *Cartographica*, 32(1): 1–21.

Sparke, M. 1998. 'A map that roared and an original atlas: Canada, cartography and the narration of nation', *Annals of the Association of American Geographers*, 88(3): 463–95.

Speier, H. 1941. 'Magic geography', *Social Researcher*, 8: 310–30.

Stark, D. and Bruszt, L. 1998. *Postsocialist Pathways: Transforming Politics and Property in East Central Europe*, New York and Cambridge, MA: Cambridge University Press.

Stengers, I. 1997. *Power and Invention: Situating Science*, Minneapolis: University of Minnesota Press.

St Martin, K. 1995. 'Changing borders, changing cartography: possibilities for intervening in the new world order', in A. Callari, S. Cullenberg and C. Biewener (eds), *Marxism in the Postmodern Age: Confronting the New World Order*, New York: The Guilford Press: 459–68.

Stone, A.R. 1995. *The War of Desire and Technology at the Close of the Mechanical Age*, Cambridge, MA: MIT Press.

Storper, M. 2000. 'Lived effects of the contemporary economy: globalization, inequality and consumer society', *Public Culture*, 12(2): 375–409.

Strategic Simulations, Inc. 1992. 'Cyber Empires', (advertisement), *Computer Game Review and CD-ROM Entertainment*, October, 2(3): 83.

Strohmayer, U. 1996. 'Pictorial symbolism in the age of innocence: material geographies at the Paris World's Fair of 1937', *Ecumene*, 3(3): 282–304.

Sugiura, N. 1982. 'Maps and the art of reasoning: the rhetorical foundations of cartographic communication', paper presented at the Annual Conference of the Association of American Geographers, San Antonio, Texas, 26 April.

Sugiura, N. 1983. 'Rhetoric and geographers' worlds: the case of spatial analysis in human geography', Ph.D. dissertation, Department of Geography, the Pennsylvania State University, State College, PA, May.

Tagg, J. 1988. *The Burden of Representation: Essays on Photographies and Histories*, Basingstoke and London: Macmillan.

Taylor, P. 1990. 'GKS', (editorial comment), *Political Geography Quarterly*, 9(3), July: 211–12.

Taylor, P. and Overton, M. 1991. 'Further thoughts on geography and GIS', (commentary), *Environment and Planning A*, 23: 1087–94.

Thomas, L.B. 1949. 'Maps as instruments of propaganda', *Surveying and Mapping*, XI(2): 75–81.

Thomsen, C.W. 1994. *Visionary Architecture: From Babylon to Virtual Reality*, New York and Munich: Prestel.

Thongchai, W. 1994. *Siam Mapped: A History of the Geo-Body of a Nation*, Honolulu: University of Hawai'i Press.

Thrower, N. 1976. 'New geographical horizons: maps', in F. Chiapelli (ed.), *First Image of America: the Impacts of the New World on the Old*, vol. 2, Berkeley and Los Angeles: University of California Press: 659–74.

Thrower, N. 1996. *Maps and Civilization: Cartography in Culture and Society*, Chicago: University of Chicago Press (1st edn, 1972).

Tobler, W.R. 1966. 'Medieval distortions: the projections of ancient maps', *Annals of the Association of American Geographers*, June 56(3): 351–60.

Tobler, W.R. 1971. 'A Cappadocian speculation', *Nature*, 231(5297), 7 May: 39–41.

Tolstoy, Leo. 1869/1986. *War and Peace*, trans. R. Edmonds, Harmondsworth: Penguin.

Tomasch, S. 1992. '*Mappae mundi* and "The Knight's Tale": the geography of power, the technology of control', in M.L. Greenberg and L. Schachterle (eds), *Literature and Technology*, Bethlehem, PA: Lehigh University Press: 66–98.

Treib, M. 1980. 'Mapping experience', *Design Quarterly*, 115.

Turnbull, D. 1993. *Maps are Territories: Science is an Atlas*, Chicago: Chicago University Press.

Turnbull, D. 1996. 'Constructing knowledge spaces and locating sites of resistance in the modern cartographic transformation', in R. Paulston (ed.), *Social Cartography: Mapping Ways of Seeing Social and Educational Change*, New York and London: Garland Publishing Inc.: 53–79.

Tyner, J.A. 1982. 'Persuasive cartography', *Journal of Geography*, July–August: 140–4.

Uebel, M. 1999. 'Toward a symptomatology of cyberporn', *Theory and Event*, 3(4), http://muse.jhu.edu/journals/theory_&_event/v003/3.4uebel.html

Ullman, E. 1953. 'Are mountains enough?', *Professional Geographer*, 5 (July): 5–8.

Ulmer, G.L. 1985. '*Applied Grammatology: Post(e)-Pedagogy from Jacques Derrida to Joseph Beuys*', Baltimore: The Johns Hopkins University Press.

Valéry, P. 1964. 'The conquest of ubiquity', *Aesthetics*, trans. R. Manheim, New York: Pantheon Books.

Vattimo, G. 1992. *The Transparent Society*, Baltimore: The Johns Hopkins University Press.

Verne, J. 1886. *Robur-l-Conquérant* (published in English in 1887 as *The Clipper of the Clouds*, London: Sampson Low).

Virilio, P. 1986. *Speed and Politics*, New York: Semiotext(e).

Virilio, P. 1997. *Open Sky*, London: Verso.

Virilio, P. and Lotringer, S. 1983. *Pure War*, New York: Semiotext(e).

Vujakovic, P. 1992. 'Mapping Europe's myths', *Geographical Magazine*, 64(9): 15–17.

Vujakovic, P. 1999. 'From "red carpet" to "iron curtain": a case study of UK mass media mapping of the European security architecture in the 1990s', paper presented at the Budapest Mass Media Map Workshop, Eotvos Lorand University, 16–19 January. http://lazarus.elte.hu/hun/tantort/1999/mmm/pv.htm

Wallas, G. 1921. *Our Social Heritage*, London: Allen & Unwin.

Waldby, C. 2000. *The Visible Human Project: Informatic Bodies and Posthuman Medicine*, New York and London: Routledge.

Wallerstein, I. (and the Gulbenkian Commission on the Restructuring of the Social Sciences). 1996a. *Open the Social Sciences: Report of the Gulbenkian Commission on the Restructuring of the Social Sciences* (Mestizo Spaces), Palo Alto: Stanford University Press.

Wallerstein, I. 1996b. 'Open the social sciences', *Items (SSRC)*, 50(1), March: 1–7.

Watts, M.J. n.d. 'Revitalizing area studies at Berkeley', http://globetrotter.berkeley.edu:80/NewGeog/

Watts, M.J. 1997. 'African studies at the *fin de siècle*: is it really the fin?', *Africa Today*, 44(2): 185–92.

Watts, M.J. 2001. '1968 and all that...', *Progress in Human Geography*, 25(2): 157–88.

Weber, S. 1987. *Institution and Intepretation* (Theory and History of Literature, vol. 31), Palo Alto: Stanford University Press.

Weigert, H.W. 1941. 'Maps are weapons', *Survey Graphic*, October: 528–30.

White, J. 1967. *The Birth and Rebirth of Pictorial Space*, Boston: Boston Book and Art Shop.

Wilson, E.O. 1998. *Consilience: The Unity of Knowledge*, New York: Knopf.

Wilson, R. and Dissanayake, W. 1996. *Global/Local. Cultural Production and the Transnational Imaginary*, Durham, NC: Duke University Press.

WITSA 2000. 'Digital planet: the global information economy', http://www.witsa. org/papers/

Wolf, A. 1991. 'What can the history of historical atlases teach? Some lessons from a century of Putsger's *Historische Schul-Atlas*', *Cartographica*, 28(2), summer: 21–37.

Wolff, L. 1994. *Inventing Eastern Europe: The Map of Civilization and the Mind of the Enlightenment*, Stanford: Stanford University Press.

Wood, D. 1977a. 'The geometry of ecstasy: more on the cartography of reality', paper presented at the Annual Meeting of the Association of American Geographers, Salt Lake City, April.

Wood, D. 1977b. 'Now and then: comparisons of ordinary Americans' symbol conventions with those of past cartographers', *Prologue*, autumn: 151–61.

Wood, D. 1978a. 'What color is the sky? An introduction to the cartography of reality', paper presented at the Annual Meeting of the Association of American Geographers, New Orleans, April.

Wood, D. 1978b. 'Shadowed spaces: in defense of indefensible space', paper prepared for the International Symposium of Selected Criminological Topics, University of Stockholm, Sweden, 11–12 August.

Wood, D. 1978c. 'Introducing the Cartography of Reality', in D. Ley and M. Samuels (eds), *Humanistic Geography: Prospects and Problems*, Chicago: Maaroufa Press: 207–19.

Wood, D. 1980. 'Night lights: data for a theory of distance', paper presented at the Annual Conference of the Association of American Geographers, Louisville, 15 April.

Wood, D. 1992. *The Power of Maps*, New York: The Guilford Press.

Wood, D. 1993. 'The fine line between mapping and mapmaking', *Cartographica*, 30(4), winter: 50–60.

Wood, D. and Fels, J. 1986. 'Designs on signs: myth and meaning in maps', *Cartographica*, 23(3): 54–103.

Woodward, D. 1992. 'Representations of the world', in R.F. Abler, M.G. Marcus and J.M. Olson (eds). *Geography's Inner Worlds: Pervasive Themes in Contemporary American Geography*, New Brunswick: Rutgers University Press: 50–73.

Woodward, D. and Lewis, G.M. 1998. *History of Cartography: Cartography in the Traditional African, American, Arctic, Australian and Pacific Societies*, vol. 2, book 3, Chicago: University of Chicago Press.

Wooley, B. 1992. *Virtual Worlds: A Journey in Hype and Hyperreality*, Cambridge, MA: Blackwell.

Wortzel, A. 1997. 'Globe theater archives: a blue planet discourse', www.intelligentagent.com/fall_globe.html 12 pp.

Wright, J.K. 1942. 'Map makers are human: comments on the subjective in maps', *Geographical Review*, 32(4) October: 527–44.

Wright, J.K. 1944. 'Human nature in science', *Science* 100 (2597), 6 October: 299–305.

Wright, J.K. 1947. '*Terrae incognitae*: the place of the imagination in geography', *Annals of the Association of American Geographers*, 37: 1–15.

Wright, J.K. 1966. *Human Nature in Geography: Fourteen Papers, 1925–1965*, Cambridge, MA: Harvard University Press.

Zelinsky, W. 1973. 'The first and last frontier of communication: the map as mystery', *Bulletin of the Geography and Map Division: Special Libraries Association*. 94: 2–8, 29.

Zizek, S. 1995. *Mapping Ideology*, London: Verso.

Zizek, S. 1999. *The Ticklish Subject*, London: Verso.

Zizek, S. 2001. *Did Somebody Say Totalitarianism?*, London: Verso.

Zuboff, S. 1984. *In the Age of the Smart Machine*, New York: Basic Books.

Index

Note: page numbers in *italics* denote illustrations

Milton Keynes UK
Ingram Content Group UK Ltd.
UKHW040106071024
449327UK00019B/840